NATIONAL GEOGRAPHIC

ANGRY BIRDS™

ANIMAL SHOWDOWN

50 Wild and Crazy Animal Face-Offs

Two Bengal tigresses (*Panthera tigris*) duke it out at an Indian zoo.

NATIONAL GEOGRAPHIC

ANGRY BIRDS™
ANIMAL SHOWDOWN

50 Wild and Crazy Animal Face-Offs

MEL WHITE
FOREWORD BY PETER VESTERBACKA

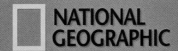

NATIONAL GEOGRAPHIC

Washington, D.C.

Published by the National Geographic Society, 1145 17th Street N.W., Washington, D.C. 20036

ISBN: 978-1-4262-1517-9 (hardcover)
ISBN: 978-1-4262-1516-2 (paperback)

The National Geographic Society is one of the world's largest nonprofit scientific and educational organizations. Its mission is to inspire people to care about the planet. Founded in 1888, the Society is member supported and offers a community for members to get closer to explorers, connect with other members, and help make a difference. The Society reaches more than 450 million people worldwide each month through *National Geographic* and other magazines; National Geographic Channel; television documentaries; music; radio; films; books; DVDs; maps; exhibitions; live events; school publishing programs; interactive media; and merchandise. National Geographic has funded more than 10,000 scientific research, conservation, and exploration projects and supports an education program promoting geographic literacy. For more information, visit www.nationalgeographic.com.

National Geographic Society
1145 17th Street N.W.
Washington, D.C. 20036-4688 U.S.A.

For information about special discounts for bulk purchases, please contact National Geographic Books Special Sales: ngspecsales@ngs.org

For rights or permissions inquiries, please contact National Geographic Books Subsidiary Rights: ngbookrights@ngs.org

Interior design: Rachael Hamm Plett

Printed in United States of America

14/QGT-CML/1

CONTENTS

Foreword: Battle Royale **6**

LEVEL 1 *ANNOYED* 8

LEVEL 2 *TESTY* 48

LEVEL 3 *OUTRAGED* 84

LEVEL 4 *FURIOUS* 120

Angry Animals by Continent **156**
Acknowledgments **158**
Illustration Credits **159**

BATTLE ROYALE

So it's not just birds that get angry!

Life on Piggy Island isn't so different from our world. The birds and the pigs may seem to be in constant conflict, but they're both just living life as they know how. Throughout our planet, animals spend their days on the lookout for food, trying to attract mates, and defending their territory—and more often than not, there's someone else about that makes their life difficult.

National Geographic has been spreading the word on how animals coexist since 1888, so they know what they're talking about when it comes to angry critters the world over.

Some of those featured in this book show that it's not the animal in the fight but the fight in the animal that counts. Most of them make the Angry Birds look pretty relaxed by comparison, and it's clear that King Pig wouldn't stand a chance in the wild!

So settle down and get stuck into some real animal showdowns! Just make sure you steer clear of those tarantula hawks!

Peter Vesterbacka
Mighty Eagle
Rovio Entertainment Ltd.

LEVEL 1 ANNOYED

(adj.) harassed, especially by quick, brief attacks

Male European hares quarrel over a female.

RAP SHEET

NAMES: JAY, JAKE, AND JIM

PHYSICAL DESCRIPTION: SMALL, ROUND, BLUE BIRDS

HOW ANGRY ARE THEY: ANNOYED

WHAT MAKES THEM ANGRY: BEING SCOLDED AND BEING IGNORED; PIGS

ANGRY BEHAVIORS: SPECIALIZING IN GROUP ATTACKS, BUT ALSO SEEN ATTACKING AS ONE

HOBBIES: MUSIC, ACTING, DRAWING

OPPONENTS: MINION PIGS, CORPO-RAL PIG, CHUCK (IN GOOD FUN!)

THREE OF A KIND

These rambunctious birds are the goofiest of the bunch, and they always work together to make even their most boring responsibilities fun and exciting. They work best as a team—it's almost like they can read each other's minds!

JUST FOR LAUGHS

While Red tries his hardest to keep them in line, Jay, Jake, and Jim (aka the Blues) love to play pranks on the other birds, especially Chuck. Their love for games and pranks gets them into trouble sometimes, though, like when it distracts them from keeping careful watch over the eggs.

Fortunately for the Blues, they are crafty enough to cover up anything that goes wrong while they've been off playing games. When the eggs have been stolen on their watch, they almost always manage to get them back before Red realizes they're missing.

THE BLUES VS. MINION PIGS

While the Blues play plenty of pranks on their fellow birds, their favorite targets are the Minion Pigs—because they're so much easier to play tricks on!

With Corporal Pig and Foreman Pig constantly barking orders at them, the Minion Pigs can usually be found trying to sneak into the birds' camp to steal the eggs. They've used all sorts of costumes and disguises, even trying to dress up like one of the Blues to get close to the nest.

Luckily for the birds, the Minion Pigs don't have the same talent for trickery as the Blues, and they've gotten caught every single time. Though they're not the most serious of birds, when it comes to protecting their family, they are definitely true blue.

AAAAAARGH!

RAP SHEET

SPECIES: VIRGINIA OPOSSUM
(DIDELPHIS VIRGINIANA)

PHYSICAL DESCRIPTION: 2.5 FT
(76 CM) LONG; 8.8-13.2 LB
(4-6 KG); FUR RANGES FROM
LIGHT GRAY TO BLACKISH

RANGE: EASTERN NORTH AMER-
ICA THROUGH CENTRAL AMERICA;
INTRODUCED TO WEST COAST OF
UNITED STATES

ANGRY BEHAVIORS: SNARLING,
BITING, PLAYING DEAD, EXUDING
FOUL-SMELLING FLUID

FORMIDABLE OPPONENTS: DOGS,
COYOTES, BOBCATS, OWLS, FOXES,
SNAKES

A young Virginia
opossum (Didel-
phis virginiana)

NOCTURNAL NIGHTMARE

The Virginia opossum is most famous for its bizarre reaction to danger. When confronted by a hungry bobcat or other enemy, the opossum may growl or hiss in anger, but it quickly goes into a coma-like state of paralysis, curled up on its side with its tongue hanging out. Yes, it fakes its own death in an attempt to confuse its attacker.

QUIT!

BETTER OFF "DEAD"

How did the opossum's odd "death" defense behavior evolve? Slow, awkward, and—let's face it—maybe not the smartest critter in the woods, an opossum is no match for a menacing bobcat, coyote, or owl. "Playing 'possum" gives it the best chance to survive a confrontation. "Why mess with me?" it seems to say. "I'm already dead."

If the strategy succeeds and the enemy leaves, the opossum eventually recovers. It may be "dead" for a few minutes or a few hours. (The process of losing and regaining consciousness is involuntary and not something the animal chooses to do.) It then stands up and makes its escape, the better to die another day.

PLAYING!

OPOSSUM VS. RATTLESNAKE

The opossum has another defense, too—in fact, one that seems like a kind of animal super-power. Studies have shown that it's immune to some poisons, including snake venom. That means if the opossum encounters a rattlesnake or copperhead on its nightly feeding trips, it can be a predator instead of a victim. Scientists believe that the species' blood contains a substance called a protease inhibitor, which renders venom ineffective. An opossum finding a dangerous snake doesn't play 'possum, but instead may catch and eat the snake without suffering a potentially deadly bite.

POSSUM!

13

Madagascar hissing cockroach (*Grompha-dorhina portentosa*)

RAP SHEET

SPECIES: MADAGASCAR HISSING COCKROACH (*GROMPHADORHINA PORTENTOSA*)

PHYSICAL DESCRIPTION: 3 IN (7.6 CM); 0.8 OZ (22.7 G); BROWN WITH BLACK HEAD

RANGE: RAIN FORESTS OF MADAGASCAR

ANGRY BEHAVIORS: HISSING AND SHOVING

FORMIDABLE OPPONENTS: OTHER COCKROACHES, CHAMELEONS, GECKOS, AND OTHER REPTILES

THIS BUG SOUNDS OFF
WHEN IT FIGHTS

The island of Madagascar may be best known for lemurs, but it's also home to lots of other odd animals, including bugs that battle in a weird and unique way. When they fight, the rain forest echoes with the fierce sound of . . . hissing.

Madagascar hissing cockroaches do indeed hiss, making the sound by forcing air through spiracles (breathing pores) on the sides of the abdomen. They hiss when they're disturbed and when they're feeling romantic; males and females hiss as part of their mating ritual.

COCKROACHES CLASH!

Males also hiss at each other when they're vying for dominance on the floor of the rain forest where they live. They fight by ramming into each other with their horns or pushing each other with their abdomens. What role does hissing play? An especially forceful hisser may scare off a lesser male and avoid fighting altogether.

Unlike many other types of roach, Madagascar hissing cockroaches have no wings and thus can't fly. Instead of laying eggs, females carry their eggs inside their bodies and give birth to young called nymphs, which then go through several instars (life stages) before becoming adults.

PESTS OR PETS?

Madagascar hissing cockroaches have become popular pets—at least among people with a high tolerance for creepiness. They do well living in a small terrarium and happily eat dry dog food and the occasional bit of rotten fruit.

Clownfish swim among a purple sea anemone.

ANEMONE IS NO ENEMY!

RAP SHEET

SPECIES: CLOWN ANEMONEFISH (AMPHIPRION OCELLARIS)

PHYSICAL DESCRIPTION: 4.3 IN (11 CM) LONG; ORANGE AND WHITE BODY

RANGE: TROPICAL PACIFIC AND INDIAN OCEANS

ANGRY BEHAVIORS: USING THE VENOMOUS SEA ANEMONE TO KEEP PREDATORS AWAY

FORMIDABLE OPPONENTS: TREVALLY, CORAL TROUT, WRASSE, AND OTHER PREDATORY FISH

PARTNERS IN SLIME

The fierce predatory fish called trevally swims tropical seas, always ready to pick off prey. The brightly colored clown anemonefish might be a nice, easy snack, but it's found a survival strategy: Make friends with something dangerous.

KEEP YOUR ANEMONES CLOSE

The clown anemonefish has developed a partnership with the sea anemone, an ocean-dwelling animal related to jellyfish that looks something like a bowl with lots of tentacles around the rim. When it is hungry, it shoots out nematocysts, which are like tiny poison-filled harpoons, to paralyze fish swimming nearby. The anemone then uses its tentacles to bring the helpless prey to its mouth for food.

ANEMONEFISH VS. TREVALLY

But the colorful clown anemonefish has figured out how to avoid becoming lunch. Beginning when it's very young, it gently and repeatedly brushes against the anemone until it's coated with the same thick, slimy mucus that covers the sea anemone. Now protected from the deadly nematocysts by a layer of slime, the clown anemonefish spends the rest of its life living amid the tentacles of the anemone unharmed.

The sea anemone protects the clown anemonefish from predators such as trevally, which don't want to suffer the stings of the nematocysts. In return, the anemonefish constantly cleans the sea anemone, picking off parasites and dead skin. The clown anemonefish also eats scraps of food from the anemone's meals—in other words, little bits of victims that strayed too close to the poison harpoons.

European
hedgehog
(*Erinaceus
europeaus*)

RAP SHEET

SPECIES: EUROPEAN HEDGEHOG (*ERINACEUS EUROPAEUS*)

PHYSICAL DESCRIPTION: 5-12 IN (13-30 CM) LONG, WITH BROWNISH SPIKES; 14-39 OZ (397-1,106 G)

RANGE: EUROPE, FROM PORTUGAL TO IRELAND AND INTO RUSSIA

ANGRY BEHAVIORS: CURLING UP TO ERECT SPINES

FORMIDABLE OPPONENTS: BADGERS, FOXES, POLECATS, OWLS

YOU WANT A FACEFUL OF SPINES?

English people love their gardens, and lots of them love hedgehogs, too. These little mammals snuffle through the grass and shrubs (and, yes, hedges) looking for food, including slugs, insects, and other invertebrates harmful to vegetables and blooming plants. Plus, hedgehogs are just so darn cute!

A PRICKLY SITUATION

They're not so cuddly, though, if you're a hungry owl, fox, or polecat (weasel). When a hedgehog feels threatened, it tucks in its head and curls up into a tight sphere with its sharp spines pointing outward, making things extremely unpleasant for any predator. Confronted by what looks like a spiky football, most enemies quickly give up and go looking for a less painful meal.

WHO YOU CALLING A HOG?

HEDGEHOG VS. BADGER

The hedgehog is in trouble in England, however. Conservationists estimate that in the past 60 years its population has fallen from 30 million to 1.5 million. Suburban sprawl and modern farming practices are part of the reason, but another cause for the loss is an increased number of badgers.

The hedgehog's defense works against many enemies, but not against the fierce badger. This large weasel uses its powerful claws and teeth to "unroll" the hedgehog's tight curl and attack its unprotected belly. This predator-prey situation is highly controversial in England, and organizations work to protect both animals. So, what's the answer? Most conservationists would say that enough habitat needs to be protected that both these species of native wildlife can find homes, coexisting in a natural balance of predator and prey in the English countryside.

Two polar bears face off in the frozen tundra.

FEISTY FACT

A POLAR BEAR'S FUR IS ACTUALLY CLEAR, NOT WHITE.

Kangaroos trade punches in Australia.

POW!

RAP SHEET

SPECIES: RED KANGAROO (*MACROPUS RUFUS*)

PHYSICAL DESCRIPTION: 2.75–5.25 FT (0.84–2 M), TAIL 2.5–4 FT (0.76–1.2 M); 80–200 LB (36–91 KG)

RANGE: MOST OF AUSTRALIA

ANGRY BEHAVIORS: "BOXING" WITH FORELEGS, KICKING WITH HIND LEGS

FORMIDABLE OPPONENTS: OTHER KANGAROOS AND, OCCASIONALLY, PEOPLE

HEAVYWEIGHT *CHAMPION* OF THE *OUTBACK*

Kangaroos can't talk (at least not in a language humans can understand), but if they could, rival males would regularly be heard calling out that old challenge, "Put up your dukes!"

KANGAROOS CLASH!

When two males ("boomers") have eyes for the same female, they settle the issue in combat that looks remarkably like a human boxing match. Standing face to face, leaning back on their long tails for support, the rivals punch each other with their forelegs, lock arms and wrestle, and, occasionally, kick at each other's bellies with their powerful hind legs—limbs so muscular they can execute jumps of 25 feet (7.6 meters) or more.

Kangaroos can do serious damage with their hind legs and feet, breaking bones or even ripping open an opponent's skin. (By the way, the species' scientific name *Macropus* means "big foot.") Most fights are less serious, though, with one challenger giving up before major injury occurs. The victor wins the favor of the female whose charms inspired the conflict.

JOGGER MEETS BOOMER

Sometimes kangaroos and humans do come into conflict. In 2010, a jogger in the Australian capital of Canberra was knocked unconscious by a kangaroo that punched the man's face without warning, giving him cuts and a black eye. Usually, though, a boomer's aggression is directed toward another of his kind, provoked by the desire for a female that stands by, watching and waiting to see which of her suitors will be victorious.

I SEE THREE LITTLE BLUE BIRDS...

GIVE PEACE A CHANCE ALREADY!

Male prairie dogs leap into battle.

RAP SHEET

SPECIES: BLACK-TAILED PRAIRIE DOG (*CYNOMYS LUDOVICIANUS*)

PHYSICAL DESCRIPTION: 12-15 IN (30-38 CM) LONG; 2-4 LB (0.91-2 KG); LIGHT-BROWN FUR

RANGE: GRASSLANDS OF CENTRAL NORTH AMERICA FROM CANADA TO MEXICO

ANGRY BEHAVIORS: KICKING AND BITING DURING THE BREEDING SEASON

FORMIDABLE OPPONENTS: RIVAL PRAIRIE DOGS, HAWKS, COYOTES

WHEN A CUTIE TURNS MEANIE

Stop for a few minutes at the viewing area near a prairie dog colony in the American West and listen to the comments of people watching. It's an absolute guarantee you'll hear the word "cute" repeatedly, along with "sweet," "adorable," and "funny."

PLAYFUL ON THE PRAIRIE

Yes, these little rodents are cute—how could you not describe them that way when group members greet each other by "kissing," when pups playfully tussle, and especially when they perform their hilarious "jump-yip" by standing up, raising their arms, and giving a sharp squeal while jumping so enthusiastically they sometimes fall over?

PRAIRIE DOGS CLASH!

But prairie dogs are far from gentle all the time. Males have territories they defend, constantly challenging neighbors with threats that sometimes lead to fights, complete with kicking and biting. And when the breeding season, or rut, arrives in early spring, they become highly aggressive, to the point that they behave completely differently from their usual playfulness.

Not only do males undergo an extreme personality change during rut, fighting nearly non-stop from dawn to dusk, this period of crazy aggression can last as long as four months. That's a lot of biting and clawing.

Some people keep prairie dogs as pets, despite an odor that might be described as pungent. These owners learn to use extreme caution during the rut, because the same robust rodent teeth adapted for gnawing can give a serious and highly painful bite if a human hand gets too close.

RAP SHEET

NAME: MATILDA

PHYSICAL DESCRIPTION: WHITE, EGG-SHAPED BIRD

ALIAS: WHITE BIRD

HOW ANGRY IS SHE: ANNOYED

WHAT MAKES HER ANGRY: DAMAGING OR DEFILING NATURE; PIGS

ANGRY BEHAVIORS: MATILDA'S PEACE-LOVING FRONT GIVES WAY TO A FIERY TEMPER WHEN SOMETHING UPSETS HER!

HOBBIES: GARDENING, BIOLOGY, AND COOKING

OPPONENTS: CHEF PIG, COOKBOOKS

DON'T STIR THE POT

For an angry bird, Matilda is pretty Zen—most of the time. She's affectionate and protective of her flock (especially the eggs), and always looks out for their best interests.

But don't let her calm look fool you! If anyone tries to touch the things Matilda cares most about—the eggs and her garden—they're sure to see exactly how angry she can get.

WHY NOT GIVE PEAS A CHANCE?

Matilda has a love for everything natural and sustainable. When she's not protecting the eggs, she's usually tending to the flowers and trees in her garden. She also uses the plants she grows to cook healthy meals for the other birds—but what she doesn't realize is that the others think her food is gross!

The other birds aren't exactly fans of acorn porridge or "salad" made from grass, but they pretend to like Matilda's cooking so her feelings won't be hurt. In her defense, she makes some pretty good herbal teas (wonderful for calming the nerves).

MATILDA VS. CHEF PIG

If there's one thing that Matilda really can't stand, though, it's when someone tries to harm her precious eggs. And that's Chef Pig's number one goal in life.

The sneaky, cunning Chef prepares all food for King Pig, who in turn wants nothing more than a big plate of scrambled eggs. Matilda finds it appalling that anyone would want to cook living things, much less something that belongs to her! She is also secretly jealous of the crafty piggy's cooking skills, but her pride would never let her admit it.

For all his cunning, Chef Pig doesn't have many allies. He's too busy scheming to make any friends, and the birds are too vigilant to let one rotten pig get to the eggs.

A cheetah races toward a Thompson's gazelle.

RAP SHEET

SPECIES: CHEETAH (ACINONYX JUBATUS)

PHYSICAL DESCRIPTION: 3.5–4.5 FT (1.1–1.4 M) LONG; TAIL 25–32 IN (64–81 CM); 77–143 LB (35–65 KG); FUR LIGHT YELLOW WITH BLACK SPOTS

RANGE: SUB-SAHARAN AFRICA, WITH AN ISOLATED POPULATION IN IRAN

ANGRY BEHAVIORS: USING SUPER BURSTS OF SPEED TO CATCH ANTELOPES

FORMIDABLE OPPONENTS: LEOPARDS, LIONS, HYENAS, ANTELOPES

ALMOST AS FAST AS ME...BUT NOT QUITE!

SPEED CAN'T BEAT STRENGTH

The cheetah is the world's fastest land mammal, able to accelerate with astounding rapidity and reach a top speed of 70 miles an hour (113 kph). From its large nostrils (for taking in air) to its big heart (for pumping blood rapidly) to its flexible spine (which lets it stretch its legs), this cat is built to go fast.

CHEETAH VS. LEOPARD

The cheetah's speed can't help it, though, when it comes to competition with more aggressive predators. Often, a cheetah that's just captured a gazelle or young zebra will have its dinner stolen by a leopard, lion, or hyena. The lithe, agile body that allows the cheetah to run so fast is no match for more powerful animals looking for an easy meal.

AFTER BEING REJECTED BY HER MOTHER, SANURRA THE CHEETAH FOUND A LIFELONG PAL IN ELLIE, AN ANATOLIAN SHEPHERD. #UNLIKELYFRIENDS

It's ironic that people often confuse the cheetah and the leopard, because the latter is one of the primary threats to young cheetahs, as well as a major thief of cheetah kills. Both are large, spotted African cats, but if the cheetah is a sprinter, the leopard is a football running back: strong enough to climb a tree while carrying an antelope.

WOLFING ITS FOOD

A cheetah will try to find a hidden spot where it can consume its food undisturbed, but its major strategy is simply to eat as much as possible as quickly as possible. In this way, it tries to finish its meal before the prey is seen or scented by a stronger animal. Even so, studies have shown that up to 50 percent of a cheetah's food is taken away by another predator.

WATCH OUT!
IT'S A TRAP!

A fly perches
on an open
Venus flytrap.

REVENGE OF THE PLANTS

As many people know only too well, animals eat plants. Countless gardeners have been dismayed to discover that caterpillars have devoured their tomato plants, or rabbits have devastated their broccoli. In fact, nature is full of plant-eating animals, from hamsters to elephants.

A few plants turn the tables, though, including one inconspicuous species that grows only in a small area of the eastern United States. It's such a special plant that the great naturalist Charles Darwin described it as "one of the most wonderful in the world": It's the Venus flytrap.

Like other carnivorous plants, including sundews and pitcherplants, the Venus flytrap grows in poor soils that lack elements such as nitrogen and phosphorus. To make up for this, these plants trap insects and spiders—and sometimes animals as large as frogs—and digest their bodies, gaining needed nutrients.

VENUS FLYTRAP VS. INSECTS

The tip of a Venus flytrap leaf forms a pair of lobes with tiny hairs that serve as triggers. When an insect walks across the leaf and touches these hairs, the lobes snap shut in only one-tenth of a second, trapping the victim. The trigger hairs are sensitive enough to distinguish between stimulation by prey and nonprey items such as falling debris and raindrops, thus saving the plant the wasted energy of trapping something it can't eat.

The lobes also allow very small insects to escape, since they wouldn't provide enough nutrition to make it worthwhile for the plant to digest them. If this happens, the leaves open again within a day to wait for a more worthwhile meal.

Continued on p. 32

Once the prey has been caught, the plant secretes various enzymes that liquefy the insect's insides. After about ten days, nothing is left of the victim but its hard shell. The rest has gone into feeding the Venus flytrap. The leaves open, the dry husk of the insect blows away in the wind, and the carnivorous plant is ready for another victim to blunder into its hungry "mouth."

AN ELECTRICALLY OPERATED TRAP

Amazingly, recent studies have indicated that the trigger hairs send a tiny electrical charge through the Venus flytrap leaf, which causes the lobes to close around the prey. Touching one hair isn't enough; two hairs must be stimulated within a few seconds for the electrical charge to activate the trap.

PLANT POACHERS

The Venus flytrap can be a victim, too. The plant is endangered by "plant rustlers" who dig it up and sell it as a houseplant. It's estimated that only about 40,000 plants are left in the wild.

RAP SHEET

SPECIES: VENUS FLYTRAP (DIONAEA MUSCIPULA)

PHYSICAL DESCRIPTION: UP TO 17 IN (43 CM) HIGH, 8 IN (20 CM) ACROSS; WHITE FLOWERS WITH FIVE PETALS

RANGE: A SMALL AREA OF NORTH AND SOUTH CAROLINA IN THE EASTERN UNITED STATES

ANGRY BEHAVIORS: LURING INSECTS AND EATING THEM

FORMIDABLE OPPONENTS: ANTS, BEETLES, SPIDERS, AND OTHER INVERTEBRATES; HUMANS

I SHOULD'VE LISTENED TO THE BIRD.

A Venus flytrap with its lunch

33

A fence lizard shows off the fire ant on its tongue.

RAP SHEET

SPECIES: FIRE ANT (SOLENOPSIS INVICTA)

PHYSICAL DESCRIPTION: 0.1 INCH (0.25 CM) LONG

RANGE: SOUTH AMERICA; INTRODUCED TO SOUTHEAST-ERN UNITED STATES

ANGRY BEHAVIORS: PAINFUL TOXIC STING

FORMIDABLE OPPONENTS: FENCE LIZARDS, HUMANS, OTHER INSECTS

SPREADS LIKE WILDFIRE

How did these little red insects get the name "fire" ants? Thousands of people in the south-eastern United States know the answer from very painful experience.

Accidentally introduced into the United States from South America in the 1930s, fire ants have spread across more than a dozen states, taking over vast areas with colonies that may have a half-million residents. They drive out other ants and use their toxic stings to kill rodents, reptiles, ground-nesting birds, and even animals as large as fawns and calves.

When their nest is disturbed, fire ants swarm out in red waves and can sting a human dozens of times in a matter of seconds, causing extreme, burning pain as well as a possible serious allergic reaction. The stings create blisters and itching that lasts for days. Some people have actually died from fire ant stings.

FIRE ANT VS. FENCE LIZARD

Amazingly, at least one animal in fire ant territory has already shown evolutionary adaptation to this relatively new danger, finding a way to win at least a few of its battles with the ants. The fence lizard is common in areas where fire ants live. When this small reptile walks into an ant colony and is attacked, it does a kind of shaking movement and runs away to rid itself of ants. Experiments have shown that individual lizards of the same species unfamiliar with fire ants remain motionless.

LONG-LEGGED LIZARD

It's not just the lizard's behavior that's changed; it's also its body shape. Fence lizards from areas where fire ants have been present for decades have adapted to have longer hind legs. These longer legs allow them to shake more vigorously and escape more quickly.

FEISTY FACT

THE COLOR RED DOESN'T MAKE COWS ANGRY—THEY CAN'T EVEN SEE IT BECAUSE THEY'RE COLOR BLIND.

A small dog and a cow face off on a farm.

A chameleon
licks up a meal.

RAP SHEET

SPECIES: MELLER'S CHAMELEON (*TRIOCEROS MELLERI*)

PHYSICAL DESCRIPTION: 21 IN (53 CM) LONG; 14.4 OZ (408 G); USUALLY GREEN AND WHITE, BUT CAN CHANGE TO BLACK, GRAY, OR BROWN

RANGE: EASTERN AFRICA

ANGRY BEHAVIORS: FIRING TONGUE LONG DISTANCES TO NAB PREY, CHANGING COLORS TO SIGNAL AGGRESSION

FORMIDABLE OPPONENTS: SMALL ANIMALS LIKE INSECTS AND OTHER LIZARDS

THE LIZARD QUICK-DRAW CHAMPION

Among the world's most bizarre reptiles, chameleons have inspired many legends to go with the facts about them—which are strange enough not to need exaggeration. A chameleon's tongue, for instance, is its deadliest weapon.

SPECIAL FEATURES

Found in Africa, Asia, and parts of Europe, chameleons are famous for being able to change color, a process that involves expanding or shrinking layers of skin that contain different pigments. It's not true, though, that they can choose any color to camouflage themselves against their background. Each species of chameleon has a limited palette that it uses to indicate aggression or willingness to mate.

Chameleons have oddly shaped toes to cling to branches. They move with a funny bouncing gait and can rotate their eyes independently of each other to see in a 360-degree field of view. Their weirdness has inspired fear at times. In the Middle East, some rural people believe that hearing a chameleon hiss will cause the listener to go blind.

CHAMELEON VS. GRASSHOPPER

Chameleons aren't dangerous, though, except to the small animals on which they prey. They have extremely long, muscular tongues (longer than their bodies) that they can shoot out with amazing speed to capture grasshoppers, mantises, butterflies, and even small birds. Firing its tongue in less than one-tenth of a second, the chameleon captures its prey with a combination of stickiness and suction, quickly drawing the victim into its mouth.

RAP SHEET

SPECIES: PLATYPUS (ORNITHO-RHYNCHUS ANATINUS)

PHYSICAL DESCRIPTION: 20 IN (51 CM) LONG, INCLUDING TAIL; 3 LB (1.4 KG); BROWN FUR

RANGE: EASTERN AUSTRALIA, INCLUDING TASMANIA

ANGRY BEHAVIORS: USING VENOMOUS SPUR DURING FIGHTS

FORMIDABLE OPPONENTS: DOGS, FOXES, MONITOR LIZARDS, HAWKS, QUOLLS

THESE GUYS KNOW HOW TO PROTECT THEIR EGGS!

Duck-billed platypus (Ornithorhyn-chus anatinus)

BIZARRE CREATURE PACKS A
SECRET WEAPON

When English scientists received the first specimens of the platypus from Australia, they thought someone was tricking them by sewing together parts of different creatures. How could an animal have the sleek body of an otter, the wide bill and webbed feet of a duck, and the paddle-like tail of a beaver?

But the platypus turned out to be real, and in time science learned even more weird facts about it—including its hidden poison-filled dagger that can kill enemies.

IT'S A BIRD! IT'S A MAMMAL!

Australia has more than its share of crazy creatures, but the platypus may be the strangest of them all.

This animal is one of only a handful of mammals that lay eggs, like a bird, instead of giving birth to live young. It hunts underwater with its eyes closed—its bill has receptors that sense faint electric impulses from prey such as worms and crustaceans. A platypus is born with teeth, but they fall out, and adults use hard mouth plates to "chew" their food. Its dense fur traps a layer of air the way a dry suit insulates a scuba diver, keeping the platypus warm even while it swims in frigid mountain streams.

PLATYPUS VS. FOX

The platypus looks furry and cuddly, but males have a secret weapon. Bony spurs on its hind legs are connected to glands that produce a powerful poison. When a male is fighting with another platypus or defending itself against an enemy, it can stab its opponent and inject the poison. Foxes and wild dogs often attack platypuses, but if they're hit by a spur they may be killed themselves. A human inflicted with a platypus jab suffers severe pain that can last days or even weeks.

Pig-tailed macaque (*Macaca nemestrina*)

RAP SHEET

SPECIES: SOUTHERN PIG-TAILED MACAQUE (*MACACA NEMESTRINA*)

PHYSICAL DESCRIPTION: 24 IN (61 CM) LONG WITH TAIL; 32 LB (15 KG); OLIVE-BROWNISH FUR; SHORT, CURLY TAIL

RANGE: SOUTHERN THAILAND AND MALAYSIA THROUGH SUMATRA AND BORNEO

ANGRY BEHAVIORS: SHAKING BRANCHES, MAKING AGGRESSIVE FACIAL EXPRESSIONS, BITING

FORMIDABLE OPPONENTS: OTHER MACAQUES

THE THUG OF THE MONKEY WORLD?

With its prominent brow and staring eyes, the male pig-tailed macaque (muh-CACK) has a look of aggression and menace. Author Nick Garbutt writes in his book *Wild Borneo* that the pig-tailed macaque has a "thuggish quality," with body language that "sends out a clear message, 'Don't mess with me!'" The targets of that message: other pig-tailed macaques.

IN THAILAND, POLICE OFFICERS FOUND AN INJURED MACAQUE, NURSED HIM BACK TO HEALTH, AND ADOPTED HIM. NOW SANTISUK IS A MEMBER OF THE FORCE. #LAWANDORDER

PIG-TAILED MACAQUES CLASH!

Pig-tailed macaques live in groups, with a dominant male that usually settles disputes. When a young male is six, it leaves its group and joins another. Newcomers often try to displace higher-ranking males, which means bloodshed. Oddly enough, one specific "kissy-face" expression is sometimes enough to make an opponent back away.

Pig-tailed macaques can be aggressive toward humans as well as other monkeys. In the oil-palm plantations that cover much of lowland Borneo, workers dread coming face-to-face with this medium-size monkey much more than they fear meeting the larger and stronger orangutan. "Don't make eye contact" is the advice often given to people who may encounter pig-tailed macaques.

WILY HUNTERS

On the other hand, pig-tailed macaques have been trained in some places to climb trees and harvest coconuts, in part because they are able to distinguish ripe from green fruits.

It all goes to show that sometimes, maybe, a thug just needs a hug.

A European weasel (left) lunges at a rabbit.

RAP SHEET

SPECIES: SHORT-TAILED WEASEL, STOAT, OR ERMINE (*MUSTELA ERMINEA*)

PHYSICAL DESCRIPTION: 12 IN (31 CM) LONG; 6 OZ (170 G); FUR CHANGES FROM BROWN IN SUMMER TO WHITE IN WINTER

RANGE: NORTH AMERICA, EUROPE, AND ASIA; INTRODUCED TO NEW ZEALAND

ANGRY BEHAVIORS: KILLING BY BITING NECK OF VICTIMS; DRINKING BLOOD

FORMIDABLE OPPONENTS: RABBITS, RODENTS, BIRDS, LIZARDS

BLOODTHIRSTY KILLER
OR SUPER-SKILLFUL HUNTER?

Whether you call it a short-tailed weasel, a stoat, or an ermine, one thing's for sure: Few hunters are more relentless than this little animal, whose fur changes from brown in summer to white in winter.

HUNT FOR SURVIVAL

With its rapid heartbeat and high metabolism, the short-tailed weasel must eat from one-fourth to one-half its weight daily just to keep from starving. That means it's constantly on the prowl, looking for mice, shrews, birds, and lizards. Fierce and efficient in pursuit, it even preys on creatures larger than itself, such as rabbits. It kills small animals by biting them on the back of the neck, severing their spinal cord.

WEASEL VS. RABBIT

Because it's never sure when its next meal might be, the ermine sometimes kills much more than it can eat at one time, which has led some to call it bloodthirsty or say that it kills for "fun." In reality, it's stockpiling food in case prey turns scarce. It doesn't help the ermine's reputation, however, that it enjoys lapping up the blood of its victims, and that when hunting is good it eats only its favorite body part: the brain.

Rabbits can run very fast and could outdistance an ermine, but they have the unfortunate habit of "freezing" when danger approaches, hoping to avoid being seen. The tactic works for some predators, but not for the ermine, which lunges and attacks—with fatal results for a rabbit. Though a quick hop from its powerful legs can sometimes help a rabbit escape, the weasel is often much quicker.

RAP SHEET

SPECIES: VIOLET BLANKET OCTO-PUS *(TREMOCTOPUS VIOLACEUS)*

PHYSICAL DESCRIPTION: FEMALES 6 FT (2 M) LONG; MALES 1 IN (2.5 CM) LONG

RANGE: TROPICAL SEAS WORLDWIDE

ANGRY BEHAVIORS: DETACHING ARMS; USING TOXIC TENTACLES TO DEFEND AGAINST PREDATORS

FORMIDABLE OPPONENTS: SHARKS AND OTHER LARGE FISH

Female blanket octopus *(Tremoctopus violaceus)*

DON'T GET TOO COZY

On a list of bizarre sea creatures, the blanket octopus would have to be near the top. In form and function, its oddness seems endless.

To begin with, the female is about six feet (two meters) long, whereas the male grows to only one inch (three centimeters). In fact, she weighs about 40,000 times more than he does. The male blanket octopus is so inconspicuous that biologists didn't even find a live one until 2002, during a research expedition on Australia's Great Barrier Reef.

MORE GADGETS THAN JAMES BOND

As the word "blanket" implies, the female sports a kind of cape that connects four of her eight arms—a translucent pink webbing that makes her look bigger and more intimidating to potential predators such as sharks. If she's attacked, she has the ability to detach one of her arms or part of her blanket to distract whatever is chasing her. (If an octopus loses an arm, is it a septopus?) Like other octopuses, she can also spray dark "ink" into the water to confuse a predator.

BLANKET OCTOPUS VS. MAN-O-WAR

Though tiny, the male blanket octopus is far from helpless. He's immune to the stinging tentacles of Portuguese man-of-war jellyfish, so when he encounters a jellyfish he tears off an armful of tentacles and uses them as a weapon against anything trying to eat him.

The male needs to stay alive so that he can mate with a female, but, unfortunately for him, that's his only role in life. Once a male finds a female and fertilizes her 100,000 eggs, he dies.

LEVEL 2 〉 TESTY

(adj.) marked by impatience or ill humor

Two devil
scorpionfish
rumble on
the reef.

RAP SHEET

NAME: BOMB

PHYSICAL DESCRIPTION: ROUND, BLACK BIRD; TURNS RED WHEN VERY ANGRY

HOW ANGRY IS HE: TESTY

WHAT MAKES HIM ANGRY: PIGS

ANGRY BEHAVIORS: IMMEDIATELY ELIMINATING THE ENEMY BY BLOWING UP IN ANGER

HOBBY: CALLIGRAPHY

OPPONENTS: PROFESSOR PIG

HAVING A BLAST

Despite his occasional . . . blowups, Bomb is generally a laid-back bird that knows how to turn most jobs into a good time. He likes to make things as easy as possible, but that sometimes means he has explosive solutions to simple problems.

Obviously, the Blues are big fans of Bomb's natural talent for finding the fun in every situation. Red, on the other hand, wishes that Bomb would apply himself more to protecting the eggs.

DETONATE AND DEFLATE

While Bomb loves to chill out on his hammock and relax, he cares just as much about the eggs as the rest of the birds. If they're in danger, Bomb will be there and ready to blow before you can say "detonate." As soon as the eggs are back in the birds' safekeeping, however, it's straight back to naptime for Bomb.

BOMB VS. PROFESSOR PIG

Professor Pig is the smartest, and the most peace-loving, of the pigs. Surprisingly for a pig, he's very much against all forms of violence and destruction, which means that Bomb definitely makes him upset. Because of his strength and unpredictable force, Bomb is bound to cause some collateral damage just by going about his everyday life.

Bomb is always causing a ruckus when he comes to the pigs' fortress to reclaim the stolen eggs, and it interrupts the peace and quiet that Professor Pig needs to work on his inventions. That may be for the best though, as all the devices and blueprints Professor Pig designs to make life easier for the pigs are often stolen and turned into egg-napping contraptions by Moustache Pig.

YEEEEHAAAAW!

RAP SHEET

SPECIES: NINE-BANDED ARMA-DILLO (*DASYPUS NOVEMCINCTUS*)

PHYSICAL DESCRIPTION: 3 FT (0.91 M) LONG, INCLUDING TAIL; 16 LB (7.3 KG); GRAY SCALY BODY

RANGE: SOUTHEASTERN UNITED STATES THROUGH MUCH OF SOUTH AMERICA

ANGRY BEHAVIORS: CURLING INTO ARMORED BALL; DIGGING PROTECTIVE BURROW

FORMIDABLE OPPONENTS: BIG CATS, COYOTES, ALLIGATORS, HUMANS

Nine-banded armadillo
(*Dasypus novemcinctus*)

ANIMALS THAT WEAR ARMOR

When danger threatens, some animals use their speed to flee; some use claws or teeth to fight; and others use camouflage to hide. The nine-banded armadillo depends on something unique to its family in the world of mammals: its protective suit of armor.

All 20 species of armadillo sport bony plates that cover much of their body and provide a degree of protection from predators. (The word *armadillo* is Spanish for "little armored one.") Only their bellies are unarmored, and therefore vulnerable, but they have a way to compensate for that.

POWERFUL ENEMIES

Even so, many "little armored ones" fall victim to predation. Although armadillos can quickly curl up to save themselves from small attackers, powerful animals such as jaguars, bears, and alligators can "unwrap" them or bite through their shells. The main threat, though, is from humans. Many armadillos are sought for food, including the giant armadillo of South America, now endangered by overhunting.

ARMADILLO VS. MOUNTAIN LION

Mountain lions are major predators of armadillos and try to sneak up on them in surprise attacks. Armadillos' eyes are very weak, so they rely on their keen noses to detect danger. When the armadillo senses a mountain lion nearby, it uses its large, powerful front claws to quickly dig a burrow or trench, wedging itself inside so that its belly is hidden. Able to reach only the hard shell, the predator, which thought it had stumbled upon an easy target, may give up in frustration and leave to search for less challenging prey. In cases like this, the potential victim becomes the victor.

RAP SHEET

SPECIES: STRIPED SKUNK (MEPHITIS MEPHITIS)

PHYSICAL DESCRIPTION: 34 IN (86 CM) LONG; 10 LB (5 KG); BLACK FUR WITH BOLD WHITE STRIPES

RANGE: MUCH OF NORTH AMERICA

ANGRY BEHAVIORS: SPRAYING NASTY-SMELLING LIQUID

FORMIDABLE OPPONENTS: WOLVES, DOGS, GREAT HORNED OWL

YUCK! SMELLS LIKE ROTTEN EGGS!

A striped skunk takes aim before spraying.

STOP, OR I'LL SPRAY YOU!

Everybody recognizes a skunk—and not just by its distinctive black-and-white stripes, but also by its unforgettable smell. Even wolves and bears steer clear of an angry skunk.

Except for a female with young, skunks are mostly solitary. They roam alone, searching for food, including fruits, insects, frogs, and small mammals and reptiles. Most animals give skunks a wide berth, but once in a while a really hungry predator might approach. And that's when the skunk prepares its chemical weapon.

When threatened, a skunk contorts its body into a U-shape, the better to see while aiming its anal glands. If the attacker doesn't heed this warning and back off, it pays the price.

THIS DOG CAN'T UNDERSTAND WHY HIS NEW PAL ISN'T INTERESTED IN HANGING OUT. #WHYCANTWEBEFRIENDS

A NASTY BLAST

A skunk can shoot its spray 12 feet (4 meters) or more, coating its victim with an oily mix of chemicals carrying an odor that lasts for days and resists most efforts to wash it off. (The chemical compounds in skunk spray are similar to those that make rotten eggs smell bad.) People who've been sprayed often burn or bury their clothes rather than try to wash them.

SKUNK VS. DOG

Few predators will approach a skunk once they see its warning posture. But man's best friend doesn't always heed warning, and tends to be curious about new visitors. In their exploration of yards and woody areas, pet dogs often wind up victims of a stinky skunk blast. When washing proves ineffective, even the most beloved of house dogs gets banished to the doghouse.

Two Japanese macaques fight in a hot spring.

FEISTY FACT

JAPANESE MACAQUES HAVE BEEN KNOWN TO THROW SNOWBALLS.

KNOCK!

IT!

OFF!

RAP SHEET

SPECIES: LLAMA (*LAMA GLAMA*)

PHYSICAL DESCRIPTION: 6 FT (2 M) TALL; 250 LB (113 KG); FUR CAN BE WHITE, BROWN, BLACK, OR VARIEGATED

RANGE: DESCENDED FROM WILD ANCESTORS IN THE ANDES OF SOUTH AMERICA

ANGRY BEHAVIORS: BITING, KICKING, SPITTING

FORMIDABLE OPPONENTS: OTHER LLAMAS, COYOTES

Two young guanacos butt heads in a meadow.

DON'T LET THE CALM DEMEANOR FOOL YOU

Most people picture llamas as the domesticated pack animals of South American villages, happily toting heavy loads in the rugged terrain and thin air of the Andes. But these camel relatives have a tough-guy side, too.

BUILT-IN BODYGUARD GENE

In recent decades, sheep ranchers have begun using llamas as guardians of their flocks. Llamas have an instinctive dislike of coyotes and are constantly on the lookout for these predators. A guard llama will place itself between the predator and the flock and, if necessary, will chase and kick the would-be attacker. No special training is needed for this guardian role. A llama introduced into a flock bonds with the sheep within a few hours or at most a week.

IF YOU SEE THIS FACE COMING TOWARD YOU, BEST TO STEP OUT OF THE WAY! #SPITTINGMAD

LLAMAS CLASH!

The species has a unique way of facing off and competing for dominance. Both male and female llamas will spit at others in their herd to claim a superior social position, or from anger. A llama can draw spit from three different stomach compartments filled with fluids of varying degrees of disgustingness. When a llama is really ticked off, it will pull from the farthest back of these compartments for a really revolting spitball. Llamas normally don't spit at people. When this happens, it's usually because the llama was treated as a pet when it was young and now considers humans to be part of its social group.

RAP SHEET

SPECIES: DINGO (CANIS LUPUS DINGO)

PHYSICAL DESCRIPTION: 5 FT (1.5 M) LONG, INCLUDING TAIL; 22–33 LB (10–15 KG)

RANGE: AUSTRALIA AND SOUTH-EASTERN ASIA

ANGRY BEHAVIORS: HUNTING IN PACKS TO KILL KANGAROOS AND DOMESTIC LIVESTOCK

FORMIDABLE OPPONENTS: SHEEP RANCHERS, KANGAROOS, WALLABIES, WATER BUFFALO

RRRRRRROWL!

Dingo (Canis lupus dingo)

60

LET THE FUR FLY

It looks like nothing more than a medium-size yellow dog—but even that observation about the dingo is controversial. For centuries people have debated the origins of Australia's iconic wild dog: Where did it come from? How did it get to Australia? Is it a wolf, or a dog, or some ancient mix?

In 2014, scientists announced that new research proved the dingo was a distinct species, a member of the dog family but separate from wolves and dogs. Furthermore, they said, pure dingoes can be many colors, not just the standard golden brown. Whether these new findings will be accepted . . . well, that's controversial.

DIN-GO, DIN-GOING, DIN-GONE

Australians have divided feelings about the dingo. To some, it's a symbol of the wild Outback, deserving protection. Many Australians have joined groups aimed at saving the dingo from persecution, believing it to be a vital part of the native ecosystem.

To many ranchers, however, it's a killer of sheep and other livestock to be eliminated whenever possible. Australia even built a fence more than 3,000 miles (4,800 kilometers) long to try to keep dingoes out of ranching areas in the southeastern part of the country.

DINGO VS. WALLABY

Dingoes are smart, resourceful, and adaptable. They often work in small packs to chase their prey until it tires or to corner it where it can't escape. Before there were sheep in Australia, they hunted various marsupials, including kangaroos and their smaller relatives, wallabies. But just because wallabies are small doesn't mean they can't put up a fight. With their powerful legs, wallabies can jump high and fast and administer a strong kick if necessary.

Two male giraffes "necking" to determine which one is stronger.

POW!

RAP SHEET

SPECIES: GIRAFFE (*GIRAFFA CAMELOPARDALIS*)

PHYSICAL DESCRIPTION: 20 FT (6 M) TALL; 1,750–2,800 LB (794–1,270 KG); COAT OF VARIED PATTERNS OF BLACK SPOTS ON LIGHT BACKGROUND

RANGE: SUB-SAHARAN AFRICA

ANGRY BEHAVIORS: "NECKING" BATTLES FOR DOMINANCE; KICKING PREDATORS

FORMIDABLE OPPONENTS: OTHER GIRAFFES, LIONS

STICKING THEIR NECKS OUT . . .
FOR A FIGHT

The world's tallest mammal, the giraffe is the skyscraper of the animal kingdom, viewing the African savannah with eyes 20 feet (6 meters) off the ground. It avoids most dangers by its sheer size. (Would you want to fight a giant?) With its long legs, it can also gallop away at speeds up to 35 miles an hour (56 kph).

Weighing well over a ton, an adult giraffe generally is threatened only by lions. Even then, a kick from a giraffe's powerful six-foot-long legs can usually discourage an attack.

REACHING FOR THE TOP SHELF

A giraffe's long neck allows it to nibble leaves high in trees that other browsing animals can't reach—especially acacia leaves, a favorite meal. Those six-foot-long (two-meter) necks? They have the same number of vertebrae—seven—as human necks. (And, believe it or not, the same as mouse necks as well.)

GIRAFFES CLASH!

Male giraffes often engage in a kind of fighting called necking. Two individuals stand close and bang their long necks together, trying to knock each other off balance. Males also butt each other with hornlike knobs, called ossicones, atop their heads.

Necking bouts can last a half hour, and occasionally result in one fighter being knocked unconscious. Only rarely, though, does any real injury result. Eventually, one of the combatants gives up and walks away. The winner is rewarded with greater social status and interest from female giraffes.

Blue poison
dart frog
(Dendrobates
azureus)

RAP SHEET

SPECIES: BLUE POISON DART FROG (*DENDROBATES AZUREUS*)

PHYSICAL DESCRIPTION: 2 IN (5 CM); YELLOW, GOLD, GREEN, OR ORANGE

RANGE: NORTHERN COLOMBIA

ANGRY BEHAVIORS: SECRETING MINUTE AMOUNTS OF DEADLY POISON FROM ITS SKIN

FORMIDABLE OPPONENTS: MONKEYS, SNAKES, LIZARDS, AND BIRDS; THE SNAKE *LIOPHIS EPINEPHELUS* IS RESISTANT TO ITS POISON

A FATAL BEAUTY

Poison dart frogs are as dangerous as they are gorgeous. Their jewel-tone colors of blue, red, yellow, green, and gold dazzle the eye, but predators know the message behind the beauty: Touch me and pay a deadly price.

DROP DEAD GORGEOUS!

While many creatures use camouflage to hide from predators, others like the poison dart frog use vivid markings to warn potential attackers that the would-be victim is dangerous. In the case of these tiny frogs from Central and South America, the defense is some of the most powerful poison on Earth.

A DEADLY DIET

Biologists still debate how these frogs become poisonous, but it's believed that they accumulate toxins in their bodies from their diet. Pet frogs without access to their natural insect food eventually lose their toxicity, and young frogs born in captivity never become poisonous unless they're returned to the wild.

POISON DART FROG VS. SNAKE

One species, the golden poison dart frog of the rain forests of Colombia, contains enough poison to kill between 10 and 20 people. Other frogs aren't quite so toxic, but still contain enough poison to be deadly. Not only is the poison dart frog toxic, it is also aggressive.

One species of snake, though, doesn't mind the poison and preys on the golden poison dart frog. The *Liophis epinephelus* (it doesn't have a common name) can survive the frog's poison. Some biologists think the snake's saliva somehow breaks down the toxins, allowing it to live.

RAP SHEET

NAME: CHUCK

PHYSICAL DESCRIPTION: TRIANGULAR YELLOW BIRD

HOW ANGRY IS HE: TESTY

WHAT MAKES HIM ANGRY: FEELING DISRESPECTED AND UNDERVALUED; PIGS

ANGRY BEHAVIORS: WHEN HE GETS ANGRY, CHUCK WHIRLS INTO HYPERSONIC CHUCK SPEED!

HOBBY: SPORTS, KARATE

OPPONENTS: CORPORAL PIG, THE BLUES

NEED FOR SPEED

It's hard to keep up with Chuck, the speediest bird in the flock. He's always zipping from one activity to the next: fighting off pigs, protecting the eggs, practicing on the beach, trying to impress Red—just hearing about it all is dizzying!

FAST, FEATHERED, AND FEISTY

In his efforts to impress the other birds (especially Red), Chuck often goes overboard; whenever he tries to show off, things tend to end in disaster. He has always been a little jealous of Bomb, who is naturally strong and doesn't have to practice at all.

Since Chuck is always trying to outdo the other birds, he gets on their nerves sometimes. But even though the Blues tease him when he messes up, all the birds know how important Chuck is to the flock. Without him as part of the team, the pigs might have gotten hold of the eggs by now.

CHUCK VS. CORPORAL PIG

You could say that Chuck and Corporal Pig have a bit in common. Both are loyal to their leaders (Red and King Pig respectively) and have a tendency to make silly mistakes with unfortunate consequences. But ultimately while Chuck often takes things a bit too far, he's got nothing on Corporal Pig. As the leader of King Pig's army, Corporal Pig is constantly trying to impress his leader by coming up with new ways to steal the birds' eggs, even if it means using the Minion Pigs as crash dummies.

When Chuck and Corporal Pig run into each other on the island, their conflicts escalate to practically nuclear levels. While Chuck's a show-off, he cares about the eggs more than anything. Corporal Pig doesn't mind what wreckage his army of minion pigs causes in their hunt for the eggs. Luckily, the rest of the birds have Chuck's back when things start to get out of hand!

An inflated puffer fish shows its spines.

RAP SHEET

SPECIES: PORCUPINEFISH (*DIODON HOLOCANTHUS*)

PHYSICAL DESCRIPTION: FROM A FEW INCHES (CM) TO 3 FT (0.9 M) LONG; USUALLY LIGHT IN COLOR WITH DARKER PATCHES

RANGE: TROPICAL OCEANS WORLDWIDE

ANGRY BEHAVIORS: INFLATING TO BECOME SPIKE-STUDDED SPHERE; CONTAINS DEADLY POISON

FORMIDABLE OPPONENTS: LARGER PREDATORY FISH; HUMAN GOURMETS

IT'S NO PARTY WHEN THIS BALLOON INFLATES

Your fellow fishes are sleek, agile, and fast, but you're awkward and slow. So how do you make sure you're not the daily special on every predator's menu?

If you're a puffer fish, you blow up like a balloon and erect the spines on your skin, turning yourself into a spiky sphere that promises pain to any predator. Oh, and you also carry one of the world's deadliest poisons in your body.

DANGEROUS DINING

Most puffer fish create a poison called tetrodotoxin, which can sicken or kill animals that eat the fish. A single puffer fish may contain enough poison to kill 30 people. In spite of this, puffer fish is a prized delicacy in Japan (where it's called fugu) and some other countries. Only chefs trained to select the nonpoisonous parts of the fish's body can legally prepare it.

PUFFER FISH VS. COBIA

A puffer fish can inflate itself to three times its normal size or more by quickly gulping down water. Spikes that usually lie flat stick out in all directions, so the puffed-up puffer fish looks like a tennis ball covered in cactus spines. Despite these defenses, the puffer fish is a tasty snack item for the fierce, six-foot-long (two-meter) fish called cobia.

With its wide-opening jaws, the cobia can bypass its mouth and engulf a puffer fish whole. Like some other species of fish, as well as sea snakes, the cobia is immune to puffer fish poison. So, even with its poison spikes, a puffer fish with the bad luck to meet a cobia is usually a goner.

BET I'M BETTER AT BLOWING UP!

DEVIL!

AAAAAAARGH!

RAP SHEET

SPECIES: TASMANIAN DEVIL (*SARCOPHILUS HARRISII*)

PHYSICAL DESCRIPTION: 20–31 IN (51–79 CM) LONG; 9–26 LB (4–12 KG); MOSTLY BLACK WITH VARIED AMOUNTS OF WHITE

RANGE: TASMANIA

ANGRY BEHAVIORS: GIVING EERIE WAILS AND SNARLS; HUNTING WOMBATS AND OTHER MAMMALS

FORMIDABLE OPPONENTS: OTHER TASMANIAN DEVILS

Tasmanian devil
(*Sarcophilus harrisii*)

ITS BITE IS AS BAD AS ITS BARK

DOWN!

Calling somebody a devil is pretty harsh, right? So how did a terrier-size creature (the world's largest carnivorous marsupial) get this satanic name?

It could be because of its seriously spooky growls and wails, or its ferocious behavior when it's upset. Tasmanian devils often feed on carrion, and when groups gather to feed on a large carcass, things can get hostile, with lots of noise and fighting for dominance.

THIS DEVIL CLEANS ITS PLATE

True to its name, this species lives only on the Australian island of Tasmania. Tasmanian devils have a large head for their body size, and their jaws are extremely powerful. When feeding, they often eat the entire body of their prey, bones and all.

The Tasmanian devil faces a serious threat to its survival in the form of a contagious cancer that has reduced the population in some areas by as much as 80 percent. Biologists are working hard to try to prevent the spread of the disease, including keeping healthy captive devils in isolation. The current wild population is estimated to be between 10,000 and 50,000.

TASMANIAN DEVILS CLASH!

UNDER!

Some male devils sport scars on their faces, evidence of battles with other males over females during mating season, or the result of bites from females that rejected their advances. Lots of devils have scars on their rumps as well. These wounds occur during group feeding, when the devils try to push others away from food with their thick-skinned backsides.

FEISTY FACT

EUROPEAN OTTERS DIG UNDERGROUND TUNNELS TO GET FROM LAND INTO WATER.

Two European
otters fight
over territory.

Bengal tiger
(*Panthera tigris*)

RAP SHEET

SPECIES: TIGER
(*PANTHERA TIGRIS*)

PHYSICAL DESCRIPTION:
UP TO 9 FT (3 M) LONG,
INCLUDING TAIL; 240–500 LB
(109–227 KG)

RANGE: INDIA AND NEPAL
THROUGH SOUTHEASTERN
ASIA

ANGRY BEHAVIORS: STALK-
ING PREY AND KILLING WITH
POWERFUL JAWS

FORMIDABLE OPPONENTS:
HUMAN POACHERS

THE *OTHER* WHITE *MEAT*

It's bad luck to be a prey species where tigers live, and it's doubly bad if you're the chef's special every day. That's the fate of wild pigs, however—they're just too tasty for their own good.

Tigers rank among the world's most magnificent and powerful predators. They use their stripes as camouflage when possible, blending into tall grass and bamboo. They approach prey with great stealth, then spring out and pounce, grasping the neck of the victim with their fearsome teeth. Unlike the average housecat, tigers don't mind getting wet, and they sometimes hide in streams or ponds to surprise prey.

FEARED AND RESPECTED

People who live near tigers are always watchful when in their habitat. Yet tourists from around the world travel to India hoping to catch a glimpse of this magnificent animal, sometimes riding elephants into tiger preserves.

Endangered wherever they live, tigers are most common in protected areas in India. In addition to habitat loss, tigers face danger from poachers.

TIGER VS. WILD BOAR

Wild pigs are common in much of the tiger's range, and they're a favored food, along with various species of deer and water buffalo.

As powerful as they are, tigers choose female and young pigs when they can, avoiding the strong mature boars with their razor-sharp tusks. Immature pigs are especially vulnerable, lacking experience in avoiding predators. Wild pigs habitually travel in groups, and a tiger will sometimes follow a herd and pick off members one by one.

Bighorn sheep
(*Ovis canadensis*)

BEST BASHER BECOMES THE BOSS

During fall in the rugged mountains of western North American, loud crashing sounds signal one of nature's most intense and violent struggles: the head-butting battles of bighorn sheep rams.

These large mammals are famed for their massive, curved horns, which can weigh 30 pounds (14 kilograms). To establish dominance, and thus win the opportunity to mate with waiting females, rams use their horns in epic tests of strength. Two males face each other, rise up on their hind legs, and rush together, colliding headfirst at 20 miles an hour (32 kph).

ADAPTATIONS FOR ALPINE AGILITY

Bighorns show amazing agility and balance as they scamper around the steep, rocky cliffs where they spend most of their time. Their acrobatic movements are aided by feet with split hooves and large eyes that provide excellent vision. When snow covers the high mountain slopes in winter, the sheep move down to sheltered valleys where they can find food.

As females prepare to give birth in spring, they usually move back up to higher elevations, searching for sites where they're safer from predators. Mountain lions are their main enemies, along with wolves, coyotes, and bears. Bighorns can walk soon after they're born, and after about a week the mothers and young congregate in herds to spend the summer grazing on alpine vegetation.

Continued on p. 78

BIGHORN SHEEP CLASH!

How can two powerful animals weighing nearly 300 pounds (136 kilograms) butt heads with such force and not suffer immediate and severe trauma? The answer lies inside their heads. Bighorns have specially adapted skulls with a double thickness of bone, as well as shock-absorbing cavities and extra-strong tendons. Most battles last only a short time, but, occasionally, two males will crash together not just once but repeatedly for hours at a time. Though injuries sometimes occur, in a great majority of contests, the weaker or less determined ram finally just gives up.

Two bighorn rams butt heads.

It usually takes at least seven years for a male to attain the body weight and horn size to join in the competition for females. Even the most powerful and dominant rams eventually get old, though, and give way to a new generation when they begin suffering defeats in the mating battles.

RAP SHEET

SPECIES: BIGHORN SHEEP (*OVIS CANADENSIS*)

PHYSICAL DESCRIPTION: 5-6 FT (1.5-2 M) LONG; 117-279 LB (53-127 KG)

RANGE: WESTERN NORTH AMERICA FROM CANADA TO MEXICO

ANGRY BEHAVIORS: BASHING HEADS TO DETERMINE DOMINANCE

FORMIDABLE OPPONENTS: OTHER RAMS, MOUNTAIN LIONS, WOLVES, COYOTES, BEARS

I SURRENDER!

RAP SHEET

SPECIES: BANDED ARCHERFISH
(*TOXOTES JACULATRIX*)

PHYSICAL DESCRIPTION: 8 IN
(20 CM) LONG; SILVERY WITH
BLACK BARS

RANGE: COASTLINES OF THE
WESTERN PACIFIC OCEAN

ANGRY BEHAVIORS: FIRING A JET
OF WATER TO KNOCK PREY INTO
THE WATER

FORMIDABLE OPPONENTS:
INSECTS AND SMALL LIZARDS

HERE, FISHY FISHY!

A SHARPSHOOTER OF INCREDIBLE SKILL

People who don't want to feel inferior to a fish should skip ahead to the next story, because this particular fish has skills that few humans master: From just below the water's surface, the archerfish can shoot down insects flying up to several feet away.

A MASTER'S DEGREE IN PHYSICS

It's not that simple, of course. The archerfish must compensate for the refraction of light, which makes objects above water appear to be in a different place than they seem to be from underwater. It also perfectly calculates where the prey will fall and swims there in less than a tenth of a second to wait, so the victim won't have time to recover and escape.

ARCHERFISH VS. INSECTS

The archerfish swims near the water's surface, watching for insects or other small animals perched on branches above. When it spots a target, it uses its powerful muscles to shoot a stream of water from its mouth as far as six feet (two meters) to knock the prey into the water, where the archerfish quickly gobbles it up. Archerfish can even alter the jet of water so that it gains power as it rises, striking with greater force than it had when it left the fish's mouth.

It's not just resting insects at risk—archerfish can learn to hit moving targets, not just stationary ones. Other archerfish have learned the skill simply by watching an individual that knows how to do it.

If there were an animal Olympics, the archerfish would be the odds-on favorite for the gold in marksmanship.

RAP SHEET

SPECIES: TARANTULA HAWK
(PEPSIS GROSSA)

PHYSICAL DESCRIPTION: 2 IN
(5 CM) LONG; BLUE-BLACK WITH
RED-BROWN WINGS

RANGE: SOUTHERN UNITED
STATES SOUTH INTO SOUTH
AMERICA

ANGRY BEHAVIORS: PARALYZING
SPIDERS WITH STING

FORMIDABLE OPPONENTS:
TARANTULAS

A tarantula hawk
faces off against
a tarantula.

NIGHT OF THE LIVING DEAD

Tarantula hawks (actually a type of wasp) are found around the world, with some particularly large and colorful species native to the southern and western United States. They have shiny blue-black bodies and brightly colored wings, varying in hue depending on the species. Only females attack tarantulas; males simply feed on flower nectar and keep watch for potential female mates.

Generally nonaggressive (except to tarantulas), these wasps will sting people if provoked. The sting of a tarantula hawk has been described as the most severe of any insect in North America, causing excruciating pain lasting about three minutes. All in all, though, that's better than what they do to their tarantula prey.

TARANTULA HAWK VS. TARANTULA

It's a horror movie in real life: A poor creature is dragged from its home and given a sting that causes instant paralysis so that it lives the rest of its life in a zombiefied coma.

The fierce attacker is a female tarantula hawk wasp. She searches for a tarantula, because the huge spider is fated to provide plenty of food for her unborn offspring.

Once the tarantula hawk has found her target and paralyzed it with her sting, she drags the helpless spider to an underground chamber, where she lays a single egg on its abdomen.

EATEN ALIVE!

When the wasp egg hatches, it begins to feed on the still liv-ing tarantula, growing for weeks until it develops enough to form a pupa. Finally, a new adult tarantula hawk emerges from the pupa and leaves the burrow—now a burial cham-ber for the dead spider that provided its nourishment.

LEVEL 3 OUTRAGED

(adj.) aroused to anger or resentment usually by some grave offense

A brown bear and a Siberian husky do battle in the snow.

RAP SHEET

NAME: MIGHTY EAGLE

PHYSICAL DESCRIPTION:
MYSTERIOUS, SELDOM SEEN
GIGANTIC BIRD OF PREY

HOW ANGRY IS HE: OUTRAGED

WHAT MAKES HIM ANGRY:
INTRUDERS, PIGS

ANGRY BEHAVIORS: TOTAL
ANNIHILATION OF ANYTHING IN
HIS WAY, TRIGGERED BY A LOVE
OF SARDINES

HOBBY: HISTORY

OPPONENTS: CHRONICLER
PIG, THE PAST

A HERO FROM THE PAST

A great warrior with a mysterious past, the Mighty Eagle is the most elusive Angry Bird. After once failing the flock, he exiled himself to a mountaintop out of shame, and he now interacts with the other birds only on very rare occasions.

THE LONE SOLDIER

When the Blues come to visit, Mighty Eagle trades them war stories for sardines (his favorite snack). Of all the birds, he can only stand the Blues, because they make him feel like a freshly hatched chick again.

Aside from visits from the Blues, the Mighty Eagle keeps to himself. Sometimes Red and Matilda try to persuade him to come back down and join the rest of the flock in protecting the eggs to restore his former glory, but he prefers his lonely roost instead.

MIGHTY EAGLE VS. CHRONICLER PIG

Chronicler Pig is guardian of the pigs' history and laws. He does his work in the Chronicle Cave, which is full of records of every era of hog history. It's convenient that he's always with the records, because Chronicler Pig has gotten very forgetful in his old age.

In his studies, Chronicler Pig once stumbled across Mighty Eagle's deepest darkest secret, the most valuable piece of information on Piggy Island. Lucky for Mighty Eagle, every time Chronicler Pig has come close to revealing his mysterious past, he gets distracted and forgets what he was about to say.

RAP SHEET

SPECIES: INDIAN GRAY MONGOOSE (*HERPESTES EDWARDSII*)

PHYSICAL DESCRIPTION: 32 IN (81 CM) LONG, INCLUDING TAIL; 4 LB (2 KG); YELLOWISH GRAY FUR

RANGE: INDIA AND ADJOINING COUNTRIES

ANGRY BEHAVIORS: USING SPEED AND AGILITY TO CAPTURE AND KILL PREY

FORMIDABLE OPPONENTS: COBRAS

TALK IT OUT, GUYS!

A mongoose faces off against a deadly cobra.

A FEARSOME SNAKE MEETS ITS MATCH

Few sights are more chilling than a cobra poised to strike. Head raised, hood spread, a large cobra can deliver enough venom in its bite to kill a human in less than a half hour.

Yet there's one animal that not only doesn't run from a cobra; it looks on the deadly serpent as a potential meal. The gray mongoose—a weasel-like mammal found on the Indian subcontinent—is renowned for its ability to kill snakes, including cobras.

MONGOOSE VS. COBRA

Mongooses use their agility and lightning-quick reflexes to jump away when a cobra strikes, darting in to bite the snake's spinal cord and disable it. Science has also uncovered a special attribute that allows mongooses to attack cobras with impunity. Mongooses are immune to the cobra's venom, a neurotoxin that quickly causes paralysis and loss of breathing in most other bite victims.

Mongooses don't specifically seek out snakes as prey. They eat a variety of foods, including birds, lizards, rodents, and amphibians. A snake is just another item on the mongoose's buffet table.

A LEGENDARY HUNTER

Generations of children have read Rudyard Kipling's story "Rikki-Tikki-Tavi," about a pet mongoose in India that saves a family from cobra attack. The story, in Kipling's 1894 collection *The Jungle Book*, was based on an old Indian folktale.

HEY! WATCH WHERE YOU'RE WALKING!

RAP SHEET

SPECIES: CRESTED PORCUPINE (*HYSTRIX CRISTATA*)

PHYSICAL DESCRIPTION: 33–46 IN (84–117 CM) LONG, INCLUDING TAIL; 12–35 LB (5–16 KG)

RANGE: MOST OF NORTH AMERICA, EXCEPT THE EASTERN UNITED STATES

ANGRY BEHAVIORS: ERECTING BARB-TIPPED QUILLS ALL OVER ITS UPPER BODY

FORMIDABLE OPPONENTS: BOBCATS, COYOTES, FISHERS

Crested porcupine
(*Hystrix cristata*)

A PRICKLY DEFENSE STRATEGY

When you're desired prey *and* slow, you need an effective defense system. The porcupine has a good one: 30,000 ways to make an attacker feel pain.

PORCUPINE VS. BOBCAT

As the second largest rodent in North America (after the beaver), the porcupine is a highly favored food for bobcats, wolves, coyotes, and other predators. To protect itself, it has evolved long, sharp quills that cover its upper body. This spiky blanket presents a formidable barrier to attack, as the quills painfully embed themselves in the body of any animal that touches the porcupine.

Each of the 30,000 quills has a barb on the tip, which makes it very hard to remove once it has penetrated skin. In fact, the shape of the barb means that over time it actually works itself deeper into flesh. Bobcats and other predators with many quills embedded around their mouths have eventually died because they could no longer eat.

A GNAWING PROBLEM

Most people enjoy seeing porcupines in the woods, but these animals can be annoying at times. Like all rodents, porcupines are addicted to gnawing things, especially wooden objects impregnated with sweat. It's not uncommon for people to find that their canoe paddles, shovel handles, and even outhouse seats have been destroyed by porcupines.

RAP SHEET

SPECIES: ALLIGATOR SNAPPING TURTLE (*MACROCHELYS TEMMINCKII*)

PHYSICAL DESCRIPTION: 26 IN (66 CM) LONG; UP TO 220 LB (100 KG)

RANGE: SOUTHEASTERN UNITED STATES

ANGRY BEHAVIORS: TRICKING FISH INTO ITS MOUTH WITH FAKE BAIT; LIGHTNING-FAST BITE

FORMIDABLE OPPONENTS: HUMANS, FISH

Alligator snapping turtle (*Macrochelys temminckii*)

DIDN'T YOU SAY THERE WAS AN EGG UP HERE?

AN A-LURING ANGLER

Down at the bottom of many lakes and rivers in the southern United States lurks one of the most formidable reptiles in North America. The alligator snapping turtle can weigh more than 200 pounds (91 kilograms) and has a sharp-pointed beak that can mangle a hand that gets too close. It can even, in fact, sever a finger.

AT HOME IN THE WATER

This turtle does slightly resemble an alligator, but it's even more aquatic than its namesake. Female alligator snapping turtles usually crawl onto dry land only to find sandy soil in which to lay their eggs, and males rarely leave the water at all.

YOU DEFINITELY DON'T WANT TO BE THIS FISH. #ANIMALSHOWDOWN

These huge turtles spend so much time underwater that their shells are often covered with algae, making them resemble rocks or sunken logs. This camouflage works well with their inventive feeding technique.

ALLIGATOR SNAPPING TURTLE VS. FISH

In its lower jaw, the alligator snapping turtle has a thin pink appendage that looks exactly like a worm. While lying on the bottom of a lake or stream, the turtle opens its jaws wide and remains motionless while slowly wiggling its fleshy mini-tongue. In the murky water, a passing bass or sunfish sees the movement and often approaches to gobble up the "worm." The fish is fast, but the turtle is faster. With a quick snap of its jaws, the alligator snapping turtle has a meal.

These turtles have become endangered in many areas because of loss of habitat and hunting for their meat, but they are now legally protected in some states.

FEISTY FACT

MALE SCARAB BEETLES WRESTLE TO IMPRESS FEMALE SCARAB BEETLES.

Two male scarab beetles
(family Lucanidae)

YOU CAN'T SIT WITH US!

An orca pops out of the water to examine a seal.

KILLER INSTINCT

When you're being pursued by a predator, one lifesaving option might be to outthink it. That is, to use trickery, or camouflage, or some other subterfuge to escape.

But what if the predator is an orca (killer whale), which not only has one of the largest brains on the planet but also uses its intelligence to dream up its own tricks? Well, you may have a problem.

ORCA VS. SEA LION

Orcas are aquatic mammals, right? (Despite the name killer "whale," orcas are actually very large dolphins.) So if you're a seal—a favorite orca snack—you climb up on a large ice floe, out of the water, where you'll be safe. The floe is too large and heavy for even a big animal like an orca to knock over.

Orcas, however, have learned to work as a team to perform an amazing feat. They swim rapidly together on the water's surface toward the floe, creating a massive wave that washes over the ice. It may not work the first time, but sooner or later a big wave will knock the seal into the sea—and then it's mealtime.

Okay, the ice floe wasn't such a great idea. What about solid ground? Sea lions sometimes herd their pups onto a beach to rest and play. In a daring maneuver, orcas can ride a wave to "surf" onto shore to snatch sea lion pups playing in shallow water. The orcas run the risk of stranding themselves, but use the retreating wave to return to deeper water.

Continued on p. 99

YOU'LL STAY IN THE WATER IF YOU KNOW WHAT'S GOOD FOR YOU!

RAP SHEET

SPECIES: ORCA OR KILLER WHALE (*ORCINUS ORCA*)

PHYSICAL DESCRIPTION: UP TO 32 FT (10 M) LONG; UP TO 6 TONS (5,443 KG); BLACK AND WHITE

RANGE: OCEANS WORLDWIDE

ANGRY BEHAVIORS: TEAMING UP TO ATTACK WHALES AND WASH SEALS OFF ICE FLOES

FORMIDABLE OPPONENTS: NO NATURAL PREDATORS

An orca beaches itself to get to young sea lions on the shore.

98

BRUTE FORCE

These actions are unusual, though. In most cases, killer whales simply nab a seal or sea lion as it swims. With their large, heavy bodies, killer whales sometime immobilize such animals by butting them or hitting them with their tails before using their large teeth to capture the prey.

Orcas travel in groups called pods of up to 40 animals. They use a highly developed and complex system of sounds to communicate and coordinate their actions. Working together, they can hunt even big marine mammals.

KEEPING IN TOUCH

Each pod uses a different dialect in its vocalizations, which allows member whales to recognize each other and remain in contact.

Orcas in different geographic regions differ slightly in appearance and behavior as well as sounds. Some populations remain in one place, for example, whereas others roam over large areas. Some groups of orcas eat only fish; others eat mainly marine mammals. Scientists think there could be more than one species of orca, but conclusive evidence is still being gathered.

RAP SHEET

SPECIES: PRAYING MANTIS (*MANTIS RELIGIOSA*)

PHYSICAL DESCRIPTION: UP TO 6 IN (15 CM) LONG; GREEN OR BROWN

RANGE: EUROPE, ASIA, AND AFRICA; INTRODUCED TO THE UNITED STATES

ANGRY BEHAVIORS: FEROCIOUS PREDATOR OF INSECTS AND VARIOUS SMALL ANIMALS

FORMIDABLE OPPONENTS: OTHER CANNIBALISTIC PRAYING MANTISES

YOU'VE GOT SOMETHING STUCK IN YOUR TEETH.

A Chinese mantis devours its recent capture.

NOT SO SAINTLY

There you are, a butterfly happily sipping nectar from a flower, and—boom!—before you know it, your head is gone. All because you didn't notice that the nearby green twig was really a praying mantis.

PRAYING MANTIS VS. BUTTERFLY

For their size, mantids (as related species are called) rank among the world's most efficient predators. They use camouflage to wait in spots where prey is likely to show up, and when a victim is within range they strike with their front legs, grasping with sharp spines that allow no escape. The capture can take less than a tenth of a second, faster than the human eye can register.

Although mantids mostly prey on grasshoppers, flies, butterflies, moths, bees, and similar insects, they have been known to capture lizards, mice, and hummingbirds. They will even eat other mantids, especially when people raise them in captivity and young hatch together in a confined space.

DECEPTIVE DESCRIPTION

The name "praying" comes from the posture of the mantis's front legs, which are usually carried in a folded position, like that of a person praying. It's often misspelled as "preying," because that's what the mantis is famous for.

The green species of praying mantis that's best known in North America is not native to the continent. It was introduced into the United States more than a century ago from Europe.

MY, WHAT BIG TEETH YOU HAVE!

RAP SHEET

SPECIES: RED-BELLIED PIRANHA (*PYGOCENTRUS NATTERERI*)

PHYSICAL DESCRIPTION: 13 IN (33 CM) LONG; 7.5 LB (3.4 KG); SILVERY-GRAY WITH RED UNDERSIDE

RANGE: MUCH OF SOUTH AMERICA

ANGRY BEHAVIORS: SWARMING IN SHOALS TO KILL AND EAT PREY

FORMIDABLE OPPONENTS: GREAT WHITE EGRET, HUMANS

Red-bellied piranha
(*Pygocentrus nattereri*)

MURDEROUS
OR MISUNDERSTOOD?

With their razor-sharp interlocking teeth and powerful jaw muscles, piranhas are capable of biting off large chunks of flesh and even cutting through bone. Hollywood movies have portrayed piranhas as able to mutilate and kill humans in a matter of minutes, attacking in packs (called shoals) with ferocious intensity.

SOUTH AMERICAN TERRORS

More than 20 species of piranhas swim through the rivers of South America. And while there have been instances of piranhas attacking people, such incidents are much rarer than legend (or cinema) would indicate. Most of them feed on other fish, crustaceans, insects, and fruits.

TEETH LIKE THESE WOULD MAKE ANY DENTIST PROUD. #FLOSSY

PIRANHA VS. GREAT WHITE EGRET

Young water birds are often piranha victims, however, especially during South America's rainy season. During this time of year, the rivers swell and crest well above the riverbanks. Birds, such as great white egrets, are forced to nest in the only available dry place—treetops. The awkward and inexperienced nestlings sometimes lose their footing and fall right into the river. If piranhas are present, an unfortunate young bird is an easy meal.

In many places, local residents swim in rivers where piranhas are common. When water levels fall, piranhas become crowded together. With food scarce, they may attack domestic animals or people with deadly results. Blood in the water and the victim's thrashing attract more piranhas to finish the job in just minutes.

RAP SHEET

NAME: TERENCE

PHYSICAL DESCRIPTION: LARGE, ROUND, DARK RED BIRD

HOW ANGRY IS HE: OUTRAGED

WHAT MAKES HIM ANGRY: PIGS, PIGS, AND PIGS

ANGRY BEHAVIORS: PIG SQUASHING, SMASHING THROUGH EVERYTHING IN HIS PATH

HOBBIES: STARING CONTESTS, SILENCE

OPPONENTS: PIGS, ANYONE IN HIS WAY

THE BIGGER THEY ARE, THE HARDER THEY FALL

All of the Angry Birds have their special talents when it comes to destruction, but nobody is quite as well-rounded as Terence. He'll appear without a sound, smashing to bits everything in his way, and then disappear with only a trail of destruction to prove he was ever even there.

YOU WON'T LIKE HIM WHEN HE'S ANGRY

Terence's favorite hobbies include crushing, smashing, and creeping up on others when they least expect it. Once, he might have cracked a smile after he demolished one of the pigs' more impressive creations, but nobody knows for sure.

He's by far the biggest bird on the island, and his extraordinary power means that the pigs don't really make him angry—they're just a minor irritation to him. His calm exterior doesn't mean he takes any prisoners though. Not even the other birds in the flock want to be near Terence when he starts throwing his weight around.

TERENCE VS. FOREMAN PIG

While most of the pigs (and a few of the birds, to be honest) are intimidated by Terence, it's Foreman Pig who really has a problem with this big bird.

Foreman Pig is responsible for building all the pigs' structures. Terence is responsible for demolishing them. Foreman Pig blames Terence more than the other birds because of Terence's sheer size and power. It's never really occurred to Foreman Pig that his machines fall apart easily because of his poor planning.

A great white shark breaches to chase a seal.

RAP SHEET

SPECIES: GREAT WHITE SHARK (*CARCHARODON CARCHARIAS*)

PHYSICAL DESCRIPTION: UP TO 20 FT (6 M) LONG; 2.5 TONS (2,268 KG) OR MORE

RANGE: TROPICAL AND TEMPERATE OCEANS WORLDWIDE

ANGRY BEHAVIORS: ATTACKING A WIDE VARIETY OF MARINE PREY

FORMIDABLE OPPONENTS: SEALS, ORCAS

LEGENDARY JAWS

Nobody who's seen the movie *Jaws* can think of great white sharks without at least a mild shudder. The film contains enough truth to cause beachgoers concern—great whites are among the planet's most impressive predators, and people are sometimes attacked as prey.

FEEDING FRENZY

Great white sharks eat all sorts of sea creatures, from turtles and birds to dolphins and fish (including other sharks). Studies have indicated that as great whites get older and larger, their appetites turn more toward mammals such as seals, sea lions, and porpoises.

FOR THIS SHARK WRANGLER, A GREAT WHITE ISN'T QUITE SO SCARY. #MANSNEWBESTFRIEND?

GREAT WHITE SHARK VS. SEAL

At places such as South Africa's Seal Island, groups of great white sharks gather to prey on fur seals, creating a noted tourist attraction. Boats take visitors offshore to watch the sharks and marvel at their amazing power to leap from the water when attacking seals from below. (Some firms bill the spectacle as "flying sharks.")

Great white sharks have an amazing ability to detect blood in the sea, sensing it from as far as three miles (five kilometers) away. Once feeding starts at a seal colony or in a school of fish, other sharks quickly gather to join in.

What chance does a fur seal have in shark-filled waters like those around Seal Island? It's more likely to survive if it swims near the sea floor. Those that swim at the surface, where they're silhouetted against the sky, end up in the ocean's most fearsome jaws.

SO, WHAT DO YOU SAY?!

RAP SHEET

SPECIES: RED FOX (*VULPES VULPES*)

PHYSICAL DESCRIPTION: UP TO 44 IN (112 CM) LONG, INCLUDING TAIL; UP TO 15 LB (7 KG); ORANGE-RED FUR WITH VARYING AMOUNTS OF WHITE AND BLACK

RANGE: NORTHERN HEMISPHERE WORLDWIDE

ANGRY BEHAVIORS: SOMETIMES KILLING DOMESTIC CHICKENS AND RAIDING GARBAGE CANS

FORMIDABLE OPPONENTS: COYOTES, EAGLES, OWLS; HUMANS HUNT THEM FOR SPORT

Red fox
(*Vulpes
vulpes*)

THE SCOUNDREL AND THE TRICKSTER

The fox has been seen as the traditional enemy of the rabbit for so long that folktales about the pair have arisen among many cultures, from African American to Native American to Louisiana Cajun to Russian. And like many traditions, there's a bit of truth in these tales.

Quite often, the fox appears as a smart but villainous fellow. He's intent on catching the rabbit, which time after time uses trickery to escape. In real life, the rabbit isn't always so lucky.

RED FOX VS. RABBIT

Rabbits, especially cottontails in the United States, are often a favorite food for red foxes. The fox creeps close to its prey, then bounds up into the air and pounces like a cat.

In England, observers have seen foxes walk right past a rabbit that "freezes" in place, which means that the fox's spotting its prey might depend more on movement that on shape and color. Is the rabbit smart, or so scared that it literally can't move? Better to say it's instinct: Freezing is a common reaction among rabbits to avoid a predator, whether a fox or a hawk. The white "cotton" tail that was so obvious suddenly disappears, confusing the predator.

A PESKY OMNIVORE

The red fox's diet includes much more than rabbits and ranges from small rodents to birds to fruit. Increasingly, foxes are moving into urban areas, where their intelligence and adaptability can cause problems. They may kill small pets, raid garbage cans, and create havoc in backyard chicken coops.

RAP SHEET

SPECIES: AFRICAN ELEPHANT (*LOXODONTA AFRICANA*)

PHYSICAL DESCRIPTION: 8.2–13 FT (2.5–4 M) HIGH AT SHOULDER; 2.5–7 TONS (2,268–6,350 KG)

RANGE: SUB-SAHARAN AFRICA

ANGRY BEHAVIORS: VIOLENTLY ATTACKING ANYTHING IN ITS VICINITY; FIGHTING FOR DOMINANCE

FORMIDABLE OPPONENTS: OTHER ELEPHANTS, RHINOCEROSES, VEHICLES

Two elephants kick up dust in the Serengeti.

RAGING BULLS

When the world's largest land animal gets mad, there's only one thing to do: get out of its way. Actually, "mad" hardly begins to describe a bull elephant in musth. When a male's in musth, his testosterone levels zoom and an odorous liquid oozes from glands on his head. The bull becomes highly, even uncontrollably, aggressive and often violent toward his fellow elephants, other animals, and even inanimate objects such as vehicles and dwellings.

ELEPHANTS CLASH!

When bull elephants fight each other, they first signal their anger by spreading their ears, raising their tusks, and swishing their trunks. They may pretend to charge and stop at the last second. When a real clash occurs, they use their tusks and can inflict serious, or even fatal, stab wounds.

A younger male will sometimes attack an older bull that has been weakened by age. While the younger elephant may succeed, thereby establishing its own dominance and increasing its mating opportunities, it also may be defeated by the more experienced veteran, gaining wisdom to perhaps win a rematch on another day.

THE MYSTERIOUS MADNESS OF MUSTH

Scientists are unsure about many aspects of musth, which normally occurs in cycles of a year or so. It seems to be related to mating, to establishment of dominance, or perhaps to assuring genetic diversity by dispersal of breeding activity.

Whatever the cause, elephant musth is nothing to take lightly. Bulls in South African national parks have literally overturned vehicles of tourists in their rage and attacked and killed animals as large as rhinoceroses.

NO TWO ZEBRAS HAVE THE SAME PATTERN OF BLACK AND WHITE STRIPES.

Rival male
zebras
face off.

Texas horned lizard
(*Phrynosoma cornutum*)

DO YOU NEED A TISSUE?

RAP SHEET

SPECIES: TEXAS HORNED LIZARD (*PHRYNOSOMA CORNUTUM*)

PHYSICAL DESCRIPTION: UP TO 4.5 IN (11 CM) LONG; LONG, SCALY BODY

RANGE: WESTERN NORTH AMERICA FROM COLORADO TO NORTHERN MEXICO

ANGRY BEHAVIORS: SQUIRTING BLOOD FROM EYES

FORMIDABLE OPPONENTS: COYOTES, BOBCATS

WHEN *BLOOD BOILS*

Animals that are small and vulnerable have developed all sorts of bizarre defenses against predators, from noxious odors to tails that break off and wiggle like snakes. So, how about a lizard that shoots squirt guns filled with blood?

The horned lizard is a round, squatty creature often mistakenly called a horned toad, though it's a reptile and not an amphibian. It eats mostly ants, flicking them up with its sticky tongue. It has to eat a lot of them, because ants aren't very nutritious; simply feeding itself takes up much of a horned lizard's time.

COPING WITH DANGER

Slower than most lizards, it might be an easy meal for hungry predators, but it has developed several ways to cope. Its body color blends in with the sandy, rocky desert terrain where it lives, making it hard to spot, and its head and back are covered with sharp spines that discourage some animals from biting. A horned lizard can also burrow into the sand and disappear with amazing speed.

HORNED LIZARD VS. COYOTE

But what if a horned lizard encounters a coyote in a spot where escape is impossible? The coyote rushes in with its sharp teeth ready, and the poor lizard is lunch, right? Not so fast: The horned lizard has a secret weapon. It can elevate the blood pressure in its head, rupture blood vessels around its eyes, and shoot out streams of bloody liquid up to 3 feet (0.9 meter).

This may well surprise and confuse a predator—hey, wouldn't it surprise you?—but the eye fluid also contains chemicals that are apparently distasteful to coyotes (common desert predators) and possibly to bobcats. So, beware, predators—this lizard is quick on the draw.

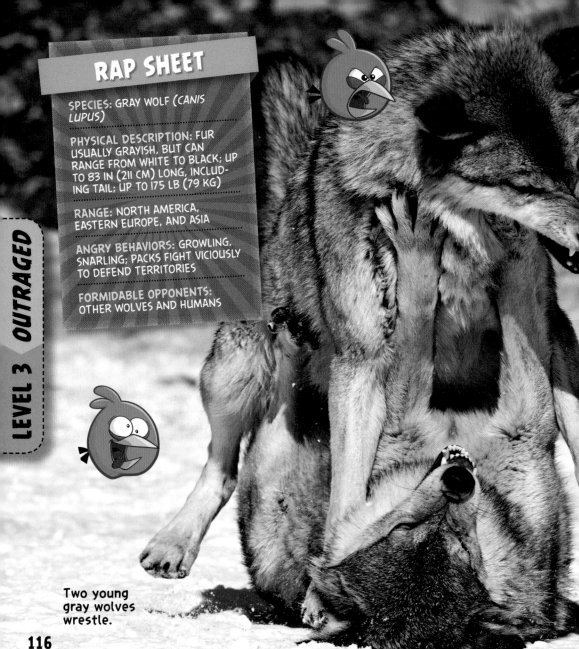

RAP SHEET

SPECIES: GRAY WOLF (*CANIS LUPUS*)

PHYSICAL DESCRIPTION: FUR USUALLY GRAYISH, BUT CAN RANGE FROM WHITE TO BLACK; UP TO 83 IN (211 CM) LONG, INCLUDING TAIL; UP TO 175 LB (79 KG)

RANGE: NORTH AMERICA, EASTERN EUROPE, AND ASIA

ANGRY BEHAVIORS: GROWLING, SNARLING; PACKS FIGHT VICIOUSLY TO DEFEND TERRITORIES

FORMIDABLE OPPONENTS: OTHER WOLVES AND HUMANS

Two young gray wolves wrestle.

116

MY, WHAT BIG TEETH YOU HAVE

The wolf looms large in the human imagination, and the image isn't always good. A fraud is a wolf in sheep's clothing; gluttons wolf their food; an unscrupulous financier is the wolf of Wall Street; Little Red Riding Hood was menaced by the Big Bad Wolf.

FAMILIES WITH FANGS

One wolf cliché isn't true, though. For decades, a wolf pack was seen as a group showing constant aggression, with members fighting for status, striving to become the "alpha" wolf. Now, biologists know that packs are usually comprised of a mated pair of wolves and their young, and dominance is merely that of parent and child.

WOLVES CLASH!

That's not to say that there isn't plenty of aggression among wild wolves. Rivalries can break out among large packs

IN FINLAND A GRAY WOLF AND A BROWN BEAR TEAMED UP TO HUNT TOGETHER. #UNLIKELYFRIENDS

that include several adult females, and fights can result in deaths. But it's territoriality that causes the great majority of wolf aggression. Packs zealously defend hunting grounds to ensure access to sufficient food for the family's survival. Territories can cover hundreds of miles, and pack members patrol constantly, watching especially for other packs intruding on their domain.

When packs meet, the result can be violent confrontations. There's plenty of growling and snarling, and teeth are bared as warring packs fight to the death over territorial boundaries. Wolf-on-wolf violence, in fact, is the leading cause of mortality in the species.

RAP SHEET

SPECIES: GOLDEN EAGLE (AQUILA CHRYSAETOS)

PHYSICAL DESCRIPTION: 33 IN (84 CM) LONG; WINGSPAN 86 IN (218 CM); 13 LB (6 KG)

RANGE: WORLDWIDE IN THE NORTHERN HEMISPHERE

ANGRY BEHAVIORS: PREYING ON MAMMALS AND BIRDS, INCLUDING SPECIES MUCH LARGER THAN ITSELF

FORMIDABLE OPPONENTS: DEER, GOATS, PRAIRIE DOGS, RABBITS

A golden eagle prepares to land.

FEATHERED FURY
FROM THE SKY

As a symbol of strength and bravery, the golden eagle is nearly unsurpassed in the animal kingdom. It served as the standard of ancient Roman military legions, and such countries as Germany, Austria, and Mexico have chosen it as their national emblem. In both French and Spanish, this species' name is "royal eagle."

The eagle's reputation is well earned, too. Though it usually feeds on small animals such as rabbits, ground squirrels, and pheasants, it's not afraid to attack far larger prey. Amazingly, golden eagles have been known to kill animals as big as deer and pronghorn.

A DEADLY RIDE

The golden eagle hunts in several ways, including diving steeply to attack waterfowl and flying close to the ground to surprise prairie dogs. To attack large mammals such as deer or goats, the eagle lands on the prey's back or neck and uses its powerful talons to pierce its victim's skin.

GOLDEN EAGLE VS. DEER

In 2011, scientists documented a golden eagle attack on a deer in eastern Russia. An automatic camera set up to capture images of Siberian tigers instead recorded the huge bird striking the deer (estimated to weigh as much as 100 pounds/45 kilograms) on its back as the victim ran and jumped in panic. The deer may have died later from injuries caused by the eagle's needle-sharp talons.

It's unknown how often this kind of predation happens, but one thing's for sure: the Roman army couldn't have chosen a more imposing symbol than the golden eagle.

I HOPE HE DIDN'T SEE ME CHECKING OUT HIS EGGS!

LEVEL 4 FURIOUS

(adj.) exhibiting ferocious displays of anger

120

Two male hippos fight for dominance.

RAP SHEET

SPECIES: BOX JELLYFISH
(*CHIRONEX FLECKERI*)

PHYSICAL DESCRIPTION:
BELL 10 IN (25 CM) ACROSS;
TENTACLES UP TO 10 FT (3 M)
LONG

RANGE: TROPICAL WATERS OF
WESTERN PACIFIC OCEAN

ANGRY BEHAVIORS: STING-
ING; INJECTS DEADLY VENOM

FORMIDABLE OPPONENTS:
SEA TURTLES, HUMANS

A school of
box jellyfish
in deep water

NOTHING SWEET ABOUT THIS JELLY

Australia is notorious for its dangerous animals, from venomous snakes to poisonous spiders to sharks. One of the least known of the country's native perils is among the deadliest of all: the box jellyfish.

A PAIN LIKE FIRE

Box jellyfish occur in near-shore waters during the Southern Hemisphere summer, from about October to June. They have tentacles up to ten feet long (three meters), but are hard to spot because of their near-transparent bodies. This combination poses a serious danger to swimmers.

Thousands of specialized stinging cells along their tentacles are capable of injecting a highly toxic poison into the jellyfish's prey, usually small fish and shrimp and other invertebrates. The poison is strong enough to kill victims quickly, which makes the box jelly deadly to humans, too. Luckily, most attacks on humans are not fatal, though they cause excruciating pain.

Unlike most jellyfish, which lack advanced vision and simply drift with the waves, box jellyfish have well-developed eyes and can choose where they swim.

BOX JELLYFISH VS. SEA TURTLE

When encountering these fearsome creatures, nearly every animal in the sea is either killed or flees in pain. The box jellyfish finds its stinging cells totally useless against the advances of a sea turtle, however.

With its hard shell and leathery skin, the sea turtle is unaffected by the box jellyfish's deadly stings. Swimming with the jellyfish's tentacles draped over its body, it happily munches the fearsome creature as a tasty snack.

HOW EVER COULD YOU EAT THAT RAW?

RAP SHEET

SPECIES: BROWN BEAR (*URSUS ARCTOS*)

PHYSICAL DESCRIPTION: UP TO 8 FT (2.4 M) TALL; UP TO 800 LB (363 KG)

RANGE: NORTHWESTERN NORTH AMERICA

ANGRY BEHAVIORS: EATING HUNDREDS OF SALMON IN LATE SUMMER

FORMIDABLE OPPONENTS: HUMANS, SALMON

A hungry grizzly bear catches salmon.

AN UPSTREAM BATTLE

Every summer in Alaska, sockeye salmon and brown bears (the coastal form of the inland grizzly bear) converge on the Brooks River with separate, and conflicting, goals: The salmon want to spawn and create a new generation of fish; the bears want to eat as many salmon as possible to survive hibernation during the long, frigid winter to come.

WHERE BEAR AND FISH MEET

Brooks Falls, in Katmai National Park, is one of the best places to see this annual spectacle of predator and prey. The waterfall creates a barrier to the fish, which have returned from the ocean and are heading upstream to spawn. For brown bears, this provides an abundant, all-you-can-eat buffet for a few weeks from July to early September.

EVEN BABY BEARS HAVE TO LEARN TO DEFEND THEMSELVES SOMETIME. #ANIMALFIGHTS

The bears wait, armed with teeth and claws. The fish have only their speed and slipperiness to help them escape the voracious predators.

BROWN BEAR VS. SOCKEYE SALMON

Salmon are a rich source of protein and fat. Bears that feed on them grow much larger than inland bears, which eat more vegetable matter such as berries and nuts. A dominant male bear occupying a prime fishing spot on the Brooks River may eat up to 30 salmon in a day. Each sockeye salmon contains about 4,500 calories, but when fish are abundant a bear may "high grade" the fish, eating only the brains, skin, and eggs.

Despite the bears' feast, thousands of salmon make it through the gauntlet to reach the spawning area. If not, there would be no more of them for bears to catch in years to come.

RAP SHEET

SPECIES: SPERM WHALE *(PHYSETER MACROCEPHALUS)*

PHYSICAL DESCRIPTION: UP TO 59 FT (18 M) LONG; UP TO 45 TONS (40,823 KG)

RANGE: OCEANS WORLDWIDE

ANGRY BEHAVIORS: FEEDING ON ENORMOUS SQUID; HISTORICAL RECORDS OF ATTACKS ON WHALING SHIPS THAT KILLED SAILORS

FORMIDABLE OPPONENTS: ORCAS, GIANT SQUID

YOU TALKIN' TO ME?

Sperm whale *(Physeter macrocephalus)*

126

WHEN DEEP-SEA GIANTS DO BATTLE

In the frigid depths of the ocean, two legendary giants meet in battle—one the hunter, the other the prey. And the most extraordinary thing about this epic encounter is that no human has actually seen it.

A LEGENDARY FIGHT

The sperm whale, the leviathan that haunted Captain Ahab in the classic novel *Moby-Dick*, roams the sea in pursuit of fish and squid. Among its favorite foods is the giant squid, a creature so mysterious that until 2004 no photograph had ever been taken of an adult in its natural habitat. With a body and tentacles that can measure more than 30 feet (9 meters), the giant squid may have inspired the mythic sea monster called the kraken.

SPERM WHALE VS. GIANT SQUID

The giant squid's eight arms and two longer tentacles are lined with round suckers, each of which is surrounded by scores of small, sharp, tooth-like spikes, which it uses to grasp and kill prey. Sperm whales captured or washed up on shore often show circular scars around their mouths from wounds caused by squid suckers. Even as the sperm whale's teeth sink into its prey, the squid fights back.

How big are these extraordinary creatures? The sperm whale has the largest brain of any animal that has ever lived, and the giant squid has the largest eyes of any living animal, to help it see in the black ocean abyss where it lives.

FEISTY FACT

THE GEMSBOK USES ITS HORNS TO FIGHT OFF LIONS AND CHEETAHS.

Two male gemsbok smash horns.

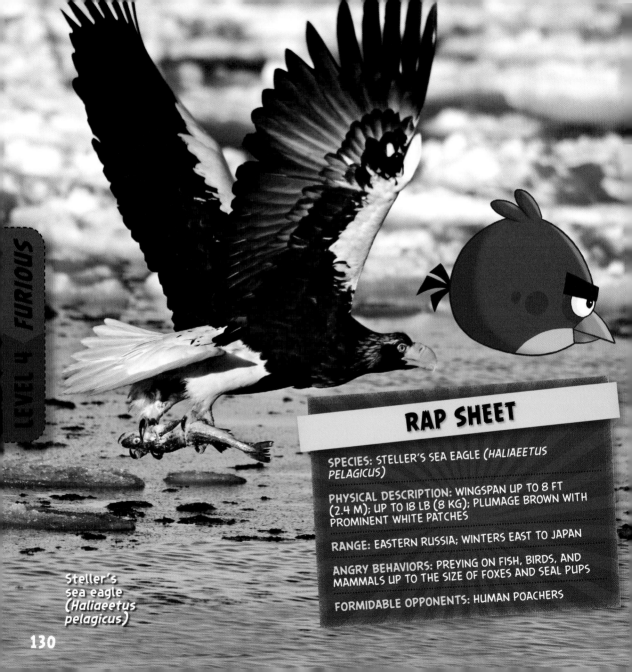

Steller's
sea eagle
(*Haliaeetus
pelagicus*)

RAP SHEET

SPECIES: STELLER'S SEA EAGLE (*HALIAEETUS PELAGICUS*)

PHYSICAL DESCRIPTION: WINGSPAN UP TO 8 FT (2.4 M); UP TO 18 LB (8 KG); PLUMAGE BROWN WITH PROMINENT WHITE PATCHES

RANGE: EASTERN RUSSIA; WINTERS EAST TO JAPAN

ANGRY BEHAVIORS: PREYING ON FISH, BIRDS, AND MAMMALS UP TO THE SIZE OF FOXES AND SEAL PUPS

FORMIDABLE OPPONENTS: HUMAN POACHERS

MY, WHAT A BIG BEAK YOU HAVE

A larger relative of the American bald eagle, the Steller's sea eagle ranks as the world's heaviest eagle—which means it's definitely one of the most formidable raptors on the planet.

NESTING INSTINCT

A pair of Steller's sea eagles will use the same nest year after year, adding branches until the nest is as much as 8 feet (2.4 meters) across, weighing several hundred pounds.

Only a few thousand Steller's sea eagles remain in the wild, and the population may be decreasing because of pollution, loss of habitat, and occasional persecution by anglers and trappers who resent its predation. Luckily, this magnificent bird is protected in both Russia and Japan, its main breeding and wintering areas. In fact, the Japanese show their respect for this species by calling it O-washi (Great Eagle).

STELLER'S SEA EAGLE VS. FISH

Like related eagles, this massive bird primarily hunts fish, catching them by zooming down to grab salmon and other large species in shallow water. Its sharp talons are well adapted to grasping slippery fish bodies, and other than swimming deeper into the water, there's not much fish can do to avoid the bird's grasp. Sometimes, an eagle even stands in the shallows and picks off fish as they swim by. Like brown bears, the Steller's sea eagle especially exploits salmon returning upriver to spawn.

The Steller's sea eagle is a powerful flier with a wingspan of up to 8 feet (2.4 meters), which means it's capable of carrying heavy prey back to its nest to feed to its chicks.

RAP SHEET

SPECIES: PEACOCK MANTIS SHRIMP (*ODONTODACTYLUS SCYLLARUS*)

PHYSICAL DESCRIPTION: UP TO 7 IN (18 CM) LONG; BODY ORANGE AND GREEN

RANGE: WESTERN PACIFIC AND INDIAN OCEANS

ANGRY BEHAVIORS: SMASHING PREY WITH SUPER-POWERFUL FORELEG STRIKE

FORMIDABLE OPPONENTS: OTHER MANTIS SHRIMP, SNAILS, HUMAN THUMBS

PEACOCK MANTIS SHRIMP (*ODONTODAC-TYLUS SCYLLARUS*)

SUPER-CRUSTACEAN KILLS WITH A POWER PUNCH

Step aside, boxing champs and karate masters—there's an ocean-dwelling animal that strikes with such force it makes your hardest jabs and punches seem like love taps.

Mantis shrimp aren't really shrimp, although as crustaceans they're related to shrimp, lobsters, and crabs. The name mantis comes from the way they fold their "raptorial appendages" (modified legs) in a way that resembles the insects called praying mantises. And like mantises, these marine animals are deadly predators.

I BET I COULD TAKE ON THIS GUY!

AN ARRAY OF WEAPONS

Hundreds of different species of mantis shrimp are found in oceans around the world. Not all punch with "clubs"; some have pointed appendages that they use to spear prey.

Many mantis shrimp are colorful, and so are sometimes sold for home aquariums. Owners must take care, though, because the largest types can break the glass walls of an aquarium or grab onto a finger. One folk name for mantis shrimp is "thumb splitter."

MANTIS SHRIMP VS. SNAIL

A mantis shrimp roams the sea floor looking for prey, with its clublike front appendages cocked and ready for use. When it approaches a victim, a unique structure allows the mantis shrimp to thrust its appendages forward like a speeding bullet, smashing through the shells of snails, crabs, and clams.

In addition to its super-powerful strike, the mantis shrimp has highly sensitive eyes that can see a much wider range of colors than the human eye. Its eyes also allow for highly accurate depth perception, which helps it aim its punches.

RRRRRRROAR!

RAP SHEET

SPECIES: MOUNTAIN GORILLA (*GORILLA GORILLA BERINGEI*)

PHYSICAL DESCRIPTION: UP TO 6 FT (1.8 M) TALL; 300–485 LB (136–220 KG)

RANGE: CENTRAL AFRICA

ANGRY BEHAVIORS: HOOTING, SNARLING, STRUTTING, CHEST BEATING

FORMIDABLE OPPONENTS: OTHER MALE GORILLAS

Mountain gorilla
(*Gorilla gorilla beringei*)

HEAR ME ROAR

A dominant silverback male mountain gorilla is the master of his troop. All the other members—younger males, adult females, and young—follow him around their home range and look to him for guidance. In return, he protects them from danger, even at the cost of his own life.

So what happens when another male shows up, maybe with the idea of challenging the local silverback for dominance? There might be a violent fight, a battle using strong arms and sharp teeth that could result in the death of one of the combatants.

MOUNTAIN GORILLAS CLASH!

It's in the best interest of individuals, and the species as a whole, to avoid unnecessary fatalities. So mountain gorillas, like many animals, have a way to settle disputes without bloodshed. A dominant silverback performs a series of actions that usually results in the challenger backing down. First, he emits off a strong odor from his armpits, signaling that he's highly upset. He may then give various hostile calls, bare his teeth, rise up on two legs, strut, and throw vegetation. In the most famous gorilla display, he beats his hands on his chest.

THIS WRESTLING MATCH BETWEEN TWO MALE MOUNTAIN GORILLAS IS ABOUT TO HEAT UP. #READYTORUMBLE

RUMBLE IN THE JUNGLE

All this is designed to make the silverback seem as intimidating as possible, and it usually succeeds in convincing the interloper to leave. But when a challenger escalates the confrontation, or when two groups meet, rival silverbacks may use their prominent canine teeth to bite and slash. Serious injury can occur or, in rare cases, death. Then, the troop could end up with a new silverback in charge.

||||||||||||||||||||||||||||||||||||||||||||||||||||||||||||||||||||||||||||||||||||||||||||||||||||

SAY CHEESE!

RAP SHEET

SPECIES: SALTWATER CROCODILE
(*CROCODYLUS POROSUS*)

PHYSICAL DESCRIPTION: UP TO
17 FT (5 M) LONG; UP TO 1,000 LB
(454 KG)

RANGE: COASTLINE OF AUSTRALIA
AND SOUTHEASTERN ASIA

ANGRY BEHAVIORS: LUNGING
FROM WATER TO GRAB VICTIMS
AND DRAGGING THEM
UNDERWATER

FORMIDABLE OPPONENTS: WATER
BUFFALO, WILD PIGS, KANGAROOS;
OCCASIONALLY, HUMANS

Saltwater crocodile
(*Crocodylus porosus*)

THE DEADLIEST
JAWS OF ALL

They look a lot like dinosaurs, and as the largest crocodilian currently living on Earth, the saltwater crocodile is one of the world's largest and deadliest predators. With jaws full of sharp teeth and the most powerful bite of any living animal, the "saltie" is one of the few predators capable of taking down anything that gets in its way.

A FEARLESS PREDATOR

Saltwater crocodiles fear nothing, and they don't hesitate to attack other large predators such as sharks, water buffalo, and even tigers.

Unfortunately, the list of croc victims sometimes includes humans. Warning signs posted on waterways across northern Australia remind beachgoers and anglers that salties are present. Still, several attacks on people occur every year in Australia and elsewhere in Asia. And with a predator as fierce as the saltwater crocodile, very few victims survive.

SALTWATER CROCODILE VS. LEOPARD

Saltwater crocs capture prey by lying quietly underwater at the edge of rivers or lakes. When an animal such as a leopard approaches the shoreline, the croc uses its powerful tail to propel itself out of the water with incredible speed and power. Grasping its victim, the croc drags it underwater, where it drowns. The prey is then either swallowed whole or, in the case of large animals such as wild pigs, deer, or kangaroos, torn apart.

RAP SHEET

SPECIES: HONEY BADGER
(MELLIVORA CAPENSIS)

PHYSICAL DESCRIPTION: UP
TO 38 IN (97 CM) LONG; UP
TO 30 LB (14 KG); BLACK FUR
WITH WHITISH UPPERSIDE

RANGE: SOUTHERN AFRICA,
MIDDLE EAST, AND INDIA

ANGRY BEHAVIORS:
DESTROYING BEEHIVES,
KILLING VENOMOUS SNAKES;
FIERCELY AGGRESSIVE

FORMIDABLE OPPONENTS:
LIONS, LEOPARDS, HUMANS

OH, IT RUNS
BACKWARD!

A honey badger
squares off with
a lion.

HONEY BADGER DON'T CARE

How tough do you have to be for *Guinness World Records* to call you "the most fearless animal in the world"? You have to be as fierce as a honey badger. Famously strong and tenacious for its size, this mammal kills deadly snakes such as cobras and puff adders and won't back down when defending its food against lions and other larger predators.

The honey badger, a weasel relative, hunts everything from scorpions to crocodiles to jackals—yet of all foods the honey badger loves bees the most.

ALMOST INDESTRUCTIBLE

It's not easy to hurt a honey badger. They have thick skin that minimizes the effect of bites, bee stings, and porcupine quills. Their skin is loose so that when a predator grabs a honey badger by the neck, it can twirl around and bite back. It's believed that honey badgers are at least partially immune to snake venom. Heavy skull bones protect them from injury and, like skunks, they can spray a vile-smelling fluid to repel attackers.

HONEY BADGER VS. PUFF ADDER

In a famous National Geographic video, a honey badger steals a gerbil from the puff adder that has just caught it. After eating the gerbil, the honey badger then kills and eats the puff adder, suffering a bite containing venom that's among the deadliest in Africa. No problem—after a few minutes of unconsciousness, the honey badger simply gets up and walks away. Now that's tough.

FEISTY FACT

THE FASTEST RECORDED SPRINTING SPEED OF A HORSE WAS 55 MPH (88 KPH).

Two wild
horses
exchange
blows.

Komodo dragon
(*Varanus komodoensis*)

AAAARGH!

RAP SHEET

SPECIES: KOMODO DRAGON
(*VARANUS KOMODOENSIS*)

PHYSICAL DESCRIPTION: UP TO
10 FT (3 M) LONG; UP TO 200 LB
(91 KG)

RANGE: LESSER SUNDA ISLANDS
OF INDONESIA

ANGRY BEHAVIORS: TEARING
PREY APART WITH FORMIDABLE
TEETH

FORMIDABLE OPPONENTS:
OTHER KOMODO DRAGONS,
WATER BUFFALO, WILD PIGS

JAWS OF DEATH

A dragonfly is fearsome only if you're a mosquito. A sea dragon threatens mostly shrimp and tiny fish. But the Komodo dragon . . . now here's a dragon that deserves its name! This powerful reptile hunts prey as large as water buffalo and wild pigs, ripping them apart with its sharp teeth and claws.

As the world's largest lizard, the Komodo dragon can grow to be ten feet (three meters) long. With its wide, scaly body; thick legs; and heavy tail, it seems a holdover from the age of dinosaurs. This powerful reptile hunts prey as large as water buffalo and wild pigs, using its sharp teeth and claws to rip apart animals much larger than itself.

HERE BE DRAGONS

The Komodo dragon was unknown to Western scientists until 1910, and today is found on only five small Indonesian islands. They're strictly protected by law, but local people sometimes kill dragons that come near villages.

KOMODO DRAGON VS. WATER BUFFALO

A Komodo dragon often hides for hours near a trail, watching for a water buffalo to approach. When the moment is right, it attacks with a rush. It can quickly kill small animals, but when the prey is as large as a water buffalo, the dragon first seriously injures it by biting. The wounded victim may travel quite a distance, while the dragon uses its highly developed sense of smell to track it. (The Komodo dragon picks up scents with its tongue, as snakes do.) The prey's death can occur hours or days later—the dragon can wait.

A vampire bat bares its fangs for the camera.

RAP SHEET

SPECIES: VAMPIRE BAT *(DESMODUS ROTUNDUS)*

PHYSICAL DESCRIPTION: 3.5 IN (9 CM) LONG; WING-SPAN 7 IN (18 CM); 2-3 OZ (57-85 G)

RANGE: MEXICO THROUGH SOUTH AMERICA

ANGRY BEHAVIORS: BITING VICTIMS AND DRINKING THEIR BLOOD

FORMIDABLE OPPONENTS: HUMANS OFTEN KILL BATS OR DISRUPT ROOSTING COLONIES

A NIGHTLY SEARCH
FOR BLOOD

Like the stars of a horror movie, these vampires wander at night, looking for victims whose blood they crave. They don't live in castles in Transylvania, though, but rather in caves in the American tropics.

Vampire bats literally live on blood. Once a young bat stops feeding on its mother's milk, it survives on nothing but blood for the rest of its life, with a body specially adapted to its unusual diet.

PEOPLE CAN BE BLOOD DONORS, TOO

Vampire bats occasionally feed on human blood. A favorite target is the exposed toe of a sleeping person. Quite often the person bitten doesn't know anything has happened until waking the next morning and finding dried blood on his or her body.

The bite from a vampire bat can become infected, but the greatest threat is rabies. Only a tiny percentage of vampire bats are rabid, but their bite has sometimes been fatal to animals and people.

VAMPIRE BAT VS. CATTLE

When night falls, the vampire bat's quest begins, as it flies out of its cave in search of food. Its target doesn't hear the beating of wings, because the bat lands near its prey, a herd of cattle, and approaches by walking. Using heat sensors on its nose, it locates a spot where blood vessels are close to the skin's surface. It then uses needle-sharp teeth to make a small, almost painless, slice in the skin. The cattle don't even know what hit them.

I CAN SHOW OFF MY TEETH TOO!

RAP SHEET

SPECIES: HIPPOPOTAMUS (*HIPPOPOTAMUS AMPHIBIUS*)

PHYSICAL DESCRIPTION: 9.5-14 FT (2.8-4 M) LONG; 2.5-4 TONS (2,268-3,629 KG)

RANGE: WETLANDS OF SUB-SAHARAN AFRICA

ANGRY BEHAVIORS: SLASHING AND BITING WITH SHARP TEETH AND POWERFUL JAWS

FORMIDABLE OPPONENTS: NILE CROCODILES, LIONS

A hippo chomps down on an attacking crocodile.

CLASH OF THE TITANS

The hippopotamus is a massive creature, weighing up to 4,000 pounds (1,800 kilograms). Despite this vegetarian diet, it has large, dagger-like teeth and muscular jaws for slashing and crushing. Combine these with its natural aggressiveness, and you have an animal that's among the most feared in Africa. (Hippos are responsible for more annual human fatalities in Africa than other feared predators such as lions and crocodiles.)

HIPPOPOTAMUS VS. NILE CROCODILE

So what happens when two of the largest, most aggressive, and most powerful animals on the planet share the same habitat? The mighty neighbors exist in a tension-filled truce that sometimes erupts into deadly conflict.

I THOUGHT THOSE GUYS WERE VEGETARIANS!

Hippopotamuses and Nile crocodiles live together along many African rivers. The croc, as the top predator in its environment, eats everything from birds and turtles to full-grown antelopes and giraffes. The hippo is a vegetarian that leaves the water nightly to graze on grass. So why doesn't the croc chomp on the occasional hippo?

DON'T MESS WITH A MAMA HIPPO

A crocodile is sometimes tempted by, and will attack, a baby hippo. But it risks the wrath of the protective mother, which won't hesitate to bite into a crocodile. While a group of crocs may successfully kill and eat a young hippo, just as often one or more of the attackers dies from hippo bites.

Lucky for the crocodiles, "live and let live" is the usual relationship between hippos and crocs. Constant violence wouldn't be a sustainable lifestyle for either.

THE FINAL SHOWDOWN

As the leader of the group, Red is one bird you definitely don't want to mess with. He's the one that found the three remaining eggs on Piggy Island, and he will do anything and everything to protect them from the pigs.

Red is big on following (and enforcing) the rules, so he's in charge of everything from waking Bomb up from a nap to keeping the Blues in line when they're goofing around. Red knows that every bird in the flock has its own special skills that are crucial to protecting the eggs before they hatch.

OH CAPTAIN, MY CAPTAIN

After discovering the three remaining eggs, Red made it his mission to keep them safe from all harm. That's not an easy task when an army of pigs is constantly coming up with new ways to steal your precious cargo. All that responsibility has shown that Red can be a true leader, as the other birds look up to him and turn to him when they need help.

The other birds sometimes get a bit frustrated that Red can never relax and enjoy himself, but he doesn't care. He has worked very hard to prevent the greedy pigs from making breakfast out of the future of the flock and doesn't want to jeopardize that.

RED VS. KING PIG

Nobody gets Red angry like King Pig—he's the one behind all of the piggy army's efforts to steal the flock's only remaining eggs, because he wants to devour them. King Pig has led the rest of the pigs into believing that he has a secret stockpile of eggs just for him.

Continued on p. 151

King Pig's deepest secret is that he's never actually eaten any eggs. He doesn't even know what they taste like! The crown on his head fools the other pigs into thinking that he's smart and powerful, but really he's just spoiled, slobbering, and lazy.

Ultimately, King Pig represents everything that Red stands against as a leader. All the responsibility that Red has had to take on results from King Pig's selfish, entitled ways. While Red would do anything to protect the eggs, sometimes he misses just hanging out with the other birds and having fun. If he could defeat King Pig once and for all, then he could go back to being just one of the flock.

RAP SHEET

NAME: RED

PHYSICAL DESCRIPTION: ROUND, RED BIRD

HOW ANGRY IS HE: FURIOUS

WHAT MAKES HIM ANGRY: ANYTHING THAT ENDANGERS THE SAFETY OF THE THREE EGGS; PIGS

ANGRY BEHAVIORS: RELENTLESSLY HURLING HIMSELF AT PIGGY FORTRESSES AGAIN AND AGAIN

HOBBIES: CHESS, PHYSICS

OPPONENTS: KING PIG, OTHER PIGS

African lion
(*Panthera leo*)

THE KING OF BEASTS TAKES ON ITS ARCHNEMESIS

In southern Africa, two predators roam the grasslands together, hunting the same prey and stealing each other's kills. One is larger and stronger; the other works in teams with the effectiveness of an attacking army.

The competition for survival between the African lion and the spotted hyena creates a continual war—always simmering, at times bursting into ferocious combat. Few rivalries in nature match species so skillful, powerful, and deadly.

WHO'S LAUGHING NOW

The lion has traditionally been depicted as a brave and proud hunter, the noble king of beasts. The hyena, on the other hand, is often seen as a skulking scavenger, cowardly and dim-witted, with a comical laughing call. Its broad head and thick neck and shoulders give it a hulking, brutish look.

In truth, the spotted hyena is a highly intelligent animal and a capable and relentless predator. Its "laugh" is just one aspect of a complex range of vocalizations that the species uses to communicate social status, danger, aggression, and other emotions. While the lion looks more imposing, the hyena is a formidable opponent. The hyena's heart is large for its body size, giving it great endurance. Teams can chase herds of a wildebeest herd for miles until one tires and succumbs to attack.

Continued on p. 154

153

I'LL LET YOU GUYS SORT THIS ONE OUT...

LION VS. SPOTTED HYENA

Hyenas scavenge other predators' kills at times, but so do lions. The two animals take over each others' kills often, fuelling their intense rivalry. Researchers have learned that lions will move toward the sounds of feeding hyenas in hopes of stealing their food. While male lions usually dominate hyenas, a group of hungry hyenas can chase a lioness away from her kill. Lions often attack and kill hyenas even when competition for food isn't involved, and hyenas will band together to chase lions away when the big cat is sufficiently outnumbered.

EVENLY MATCHED RIVALS

While the lion is undoubtedly the king of its domain, the spotted hyena plays an equally important role on the African plains. Both are dominant predators, well adapted to a life of hunting and scavenging. When they compete, the winner often depends more on circumstances than bravery.

RAP SHEET

SPECIES: AFRICAN LION *(PANTHERA LEO)*

PHYSICAL DESCRIPTION: UP TO 9.5 FT (3 M) LONG, INCLUDING TAIL; 265–420 LB (120–191 KG)

RANGE: SUB-SAHARAN AFRICA

ANGRY BEHAVIORS: STEALING KILLS FROM HYENAS; TOP PREDATOR IN ITS RANGE

FORMIDABLE OPPONENTS: SPOTTED HYENAS, HIPPOS, GIRAFFES

A male lion chases off his rival, a spotted hyena.

Angry Animals by Continent

Africa
Madagascar hissing cockroach
Cheetah
Chameleon
Giraffe
Mountain gorilla
Elephant
Hippo
Honey badger
Lion

Asia
Pig-tailed macaque
Mongoose
Tiger/wild boar
Golden eagle
Steller's sea eagle
Komodo dragon

Europe
Weasel
Red fox
Hedgehog

North America
Opossum
Prairie dog
Venus flytrap
Fire ant
Skunk
Alligator snapping turtle
Bighorn sheep
Porcupine
Praying mantis
Horned lizard
Gray wolf
Grizzly bear

South America
Armadillo
Llama
Poison dart frog
Tarantula hawk
Red-bellied piranha
Vampire bat

Australia
Kangaroo
Dingo
Tasmanian devil
Archerfish
Saltwater crocodile
Platypus

Oceans
Clownfish
Blanket octopus
Orca
Puffer fish
Great white shark
Box jellyfish
Mantis shrimp
Sperm whale

Acknowledgments

We would like to extend our thanks to the terrific team who worked so hard to make this project come together so quickly and so well.

Rovio
Laura Nevanlinna, Jan Schulte-Tigges, Rollo de Walden, and Ilona Lindh

National Geographic
Michelle Cassidy, Bridget A. English, Rachael Hamm-Plett, Moduza Design, Jonathan Halling, Karen Matthes, Matt Propert, Kristin Sladen, Galen Young, Lisa A. Walker, Katie Olsen, Judith Klein, and Marshall Kiker

Illustration Credits